MW01533209

Proceedings of the U.S. Geological Survey Eighth Biennial Geographic Information Science Workshop and First *The National Map* Users Conference, Denver, Colorado, May 10–13, 2011

Edited by Jennifer B. Sieverling and Jeffrey Dietterle

Scientific Investigations Report 2011–5053

U.S. Department of the Interior
U.S. Geological Survey

U.S. Department of the Interior
KEN SALAZAR, Secretary

U.S. Geological Survey
Marcia K. McNutt, Director

U.S. Geological Survey, Reston, Virginia: 2011

For more information on the USGS—the Federal source for science about the Earth, its natural and living resources, natural hazards, and the environment, visit http://www.usgs.gov or call 1–888–ASK–USGS.

For an overview of USGS information products, including maps, imagery, and publications, visit http://www.usgs.gov/pubprod

To order this and other USGS information products, visit http://store.usgs.gov

Suggested citation:
Sieverling, J.B., and Dietterle, Jeffrey, eds., 2011, Proceedings of the U.S. Geological Survey Eighth Biennial Geographic Information Science Workshop and First *The National Map* Users Conference, Denver, Colorado, May 10–13, 2011: U.S. Geological Survey Scientific Investigations Report 2011–5053, 91 p.

ISBN 978 1 4113 3122 8

Contents

Proceedings of the U.S. Geological Survey Eighth Biennial Geographic Information Science Workshop and First *The National Map* Users Conference, Denver, Colorado, May 10–13, 2011

Edited by Jennifer B. Sieverling[1] and Jeffrey Dietterle[2]

Introduction

The U.S. Geological Survey (USGS) is sponsoring the first *The National Map* Users Conference in conjunction with the eighth biennial Geographic Information Science (GIS) Workshop on May 10-13, 2011, in Lakewood, Colorado. The GIS Workshop will be held at the USGS National Training Center, located on the Denver Federal Center, Lakewood, Colorado, May 10-11. *The National Map* Users Conference will be held directly after the GIS Workshop at the Denver Marriott West, a convention hotel in the Lakewood, Colorado area, May 12-13.

The National Map is designed to serve the Nation by providing geographic data and knowledge for government, industry, and public uses. The goal of *The National Map* Users Conference is to enhance communications and collaboration among the communities of users of and contributors to *The National Map*, including USGS, Department of the Interior, and other government GIS specialists and scientists, as well as the broader geospatial community. The USGS National Geospatial Program intends the conference to serve as a forum to engage users and more fully discover and meet their needs for the products and services of *The National Map*.

The goal of the GIS Workshop is to promote advancement of GIS and related technologies and concepts as well as the sharing of GIS knowledge within the USGS GIS community. This collaborative opportunity for multi-disciplinary GIS and associated professionals will allow attendees to present and discuss a wide variety of geospatial-related topics.

The Users Conference and Workshop collaboration will bring together scientists, managers, and data users who, through presentations, posters, seminars, workshops, and informal gatherings, will share accomplishments and progress on a variety of geospatial topics. During this joint event, attendees will have the opportunity to present or demonstrate their work; to develop their knowledge by attending hands-on workshops, seminars, and presentations given by professionals from USGS and other Federal Agencies, GIS related companies, and academia; and to network with other professionals to develop collaborative opportunities.

Specific conference topics include scientific and modeling applications using *The National Map*, opportunities for partnerships, and advances in geospatial technologies.

The first part of the week will be the GIS Workshop, offered as a pre-conference seminar. It will focus on hands-on GIS training and seminars concerning current topics of geospatial interest. The focus of the USGS GIS Workshop is to showcase specific techniques and concepts for using GIS in support of science. The presentations will be educational and not a marketing endeavor. To promote awareness of and interaction with selected USGS corporate and local science center data products, as well as promoting collaboration, a "GIS Olympics" event will be held Tuesday evening during the GIS Workshop.

The second part of the week will feature interactive briefings and discussions on issues and opportunities of *The National Map*. The focus of the Users Conference will be on the role of *The National Map* in supporting science initiatives, emergency response, land and wildlife management, and other activities. All presentations at the Users Conference include use or innovations related to a *The National Map* data theme or application. On Wednesday evening, a poster session is being held as a combined event for all attendees and as a juncture between the events. On Thursday evening, the Henry Gannett Award will be presented. Additionally, poster awards will be presented.

[1]U.S. Geological Survey, Denver, Colorado.

[2]U.S. Geological Survey, Reston, Virginia.

Several prominent speakers are featured at plenary sessions at *The National Map* Users Conference, including Deanna A. Archuleta, Deputy Assistant Secretary for Water and Science, Department of the Interior; Dr. Barbara P. Buttenfield, Professor of Geography at the University of Colorado in Boulder; best-selling author Frederick Reuss; and Dr. Joel Scheraga, Senior Advisor for Climate Adaptation, U.S. Environmental Protection Agency. Additionally, panel discussions have attracted participation from notable experts from government, academia, and the private sector.

This Proceedings volume will serve as an activity reference for workshop attendees, as well as an archive of technical abstracts presented at the workshop. Author, co-author, and presenter names, affiliations, and contact information are listed with presentation titles with the abstracts. Some hands-on sessions are offered twice; in these instances, abstracts submitted for publication are presented in the proceedings on both days on which they are offered.

Special Events

GIS Olympics

A "GIS Olympics" event will be held Tuesday evening at the hotel venue to promote awareness of and interaction with selected USGS corporate and local science center data products, as well as promoting collaboration. More details are given in the abstract for the event.

Poster Session

A poster session will be held on Wednesday evening from 6:00 p.m. to 8:00 p.m., and awards for various categories will be presented at the Thursday evening awards session. The poster session will serve as a juncture between the two events and give attendees from both events a chance to share their work with each other. Descriptions of some of the posters are included in this publication; other posters, for which titles and abstracts were not submitted, will be presented.

The 2011 National Hydrography Dataset (NHD) Workshop

The last NHD Stewardship Conference was conducted in April 2009. The next installment of this popular event will take place at the GIS Workshop and *The National Map* Users Conference.

Henry Gannett Award Ceremony

If you have ever used a topographic map to find your way around a remote part of the country, or if you have ever taken note of how geographic names reflect the history of the land and the culture of its inhabitants, you will appreciate the pioneering work of Henry Gannett, an early American geographer often considered to be the father of topographic mapping in the United States.

The Henry Gannett Award commemorates Gannett's varied contributions and passions for American geography and cartography while recognizing sustained and distinguished contributions to contemporary USGS topographic mapping.

Born in Bath, Maine, in 1846, and educated at Harvard University, Gannett began his career in topographic mapping with the Hayden Survey in 1871. The USGS Geography Program was established under his direction, and he served as Chief Geographer of the Survey from 1882 to 1914. Under his leadership, the program's first topographic map sheets were produced. Through his work as a geographer of the U.S. censuses of 1880, 1890, and 1900, and the Philippine, Cuban, and Puerto Rican censuses, Gannett became interested in place names. His efforts to resolve difficulties caused by the confusion and duplication of geographic names, especially in Western lands, contributed to the establishment of the U.S. Board on Geographic Names in 1890. Gannett was also one of the founders of the National Geographic Society (founding member, 1883; president, 1910-14), the Geological Society of America, and the Association of American Geographers.

The National Map "Mashathon"

The National Map Users Conference will feature a "mashathon" for developers, managers, users, etc., to showcase *The National Map* data, products, services, and viewers in use. A "mashathon" is a digital map gallery. Submittals can highlight work that is published or in progress and include working demonstrations or screenshots of maps or applications in *The National Map* Viewer or other tools.

User Feedback to *The National Map*

User input is needed to shape *The National Map*. Feedback will be gathered in real-time at the conference and reviewed and discussed with the participants at the closing session. Opportunities to share ideas include the following events: Communities of Use Panel, two Listening Sessions, two Product and Services Feedback Sessions, *The National Map* Doctor's Office, onsite 'Tweeting,' and an old-fashioned suggestion box.

National Meetings

The National Digital Orthophoto Program (NDOP) Meeting and the U.S. Board on Geographic Names will meet in conjunction with the GIS Workshop and *The National Map* Users Conference. The meetings are open to conference attendees.

Participating Organizations

Federal Departments and Agencies

Bureau of Land Management
Bureau of the Census
Centers for Disease Control and Prevention
Department of Homeland Security
Department of the Interior
FEMA Region VIII
National Geodetic Survey, National Oceanic and Atmospheric Administration
National Geospatial-Intelligence Agency
Natural Resources Conservation Service
U.S. Department of Agriculture, Animal and Plant Health Inspection Service, Centers for Epidemiology and Animal Health
U.S. Environmental Protection Agency
U.S. Forest Service
U.S. Geological Survey

State Agencies

Alabama Department of Economic and Community Affairs
Alaska Department of Fish and Game
Colorado Division of Water Resources
Illinois State Geological Survey
Indiana Geographic Information Council
Kentucky Geological Survey
Louisiana Department of Transportation and Development
New Jersey Department of Environmental Protection
New Jersey Office of Information Technology
Oregon Water Resources Department
State of Utah, Automated Geographic Reference Center
Utah State Geological Survey
Wisconsin State Cartographers Office

Academia

Louisiana State University
Penn State University
University of Alaska at Fairbanks
University of Colorado
University of Idaho
University of NIS, Serbia
University of North Dakota School of Medicine
University of Texas at Dallas
University of Washington

University of Wisconsin – Madison
University of Wyoming, Wyoming Geographic Information Science Center

Commercial

Aquaveo
Altru Health System
Boundary Solutions, Inc.
Booz Allen Hamilton
Critigen
Data East, LLC
Digital Globe, Inc.
Esri
GDM International Services
Geosoft Inc.
Horizon Systems, Inc.
ITT Visual Information Solutions
Northrop Grumman
Parallel, Inc.
Photo Science
RockyMountainGeo
Sanborn Map Company
Science Applications International Corporation
TerraCarto, LLC
TerraGo Technologies
Xentity LLC

Other

Lane Council of Governments
International Joint Commission of Canada and the U.S.
Denver Regional Council of Governments

U.S. Geological Survey Geographic Information Science Workshop Coordinator

Jennifer Sieverling

Planning Committee Members

Stephen Char	Curtis Price
Catherine Costello	Barbara Ray
Jacque Fahsholtz	Carma San Juan
Wm. Steve Helterbrand	Sharon Shin
Tim McKinney	Jay Storey
Mike Mulligan	Shane Wright
Daniel Pearson	

The National Map Users Conference Coordinator

Jeff Dietterle

Planning Committee Members

Stephen Aichele	Vicki Lukas
Ariel Bates	JP Maxwell
Katrina Burke	Lindsay Murphy
William Carswell	Mark Newell
Doug Clark	Patricia Phillips
Catherine Costello	Barbara Ray
Kari Craun	Jennifer Sieverling
Tracy Fuller	Jeff Simley
Bruce Geyman	E. Lynn Usery
Wm. Steve Helterbrand	Carl Zulick
Patricia Hytes	

Plenary Session Speakers

Kevin T. Gallagher – Master of Ceremonies
Associate Director, Core Science Systems
U.S. Geological Survey

Mr. Gallagher serves as the Associate Director, Core Science Systems, and oversees the USGS Geologic Mapping, Geological and Geophysical Data Preservation, Geospatial, Biological Information, and Science Informatics Programs as well as the world's largest Earth Science Library.

From 2002 to 2010, Mr. Gallagher served as the USGS Chief Information Officer and Chief Technology Officer, where he oversaw the operation of information technology systems and networks supporting bureauwide computing and telecommunications.

Before joining the USGS, Mr. Gallagher held a number of information technology and management positions at various Federal agencies, including Chief, Operations Division, U.S. Coast Guard Operations Systems Center, where he oversaw the development and operations of computer systems supporting Search and Rescue, Environmental Protection, Marine Safety, and Law Enforcement; and Software Developer with the Department of the Navy and Naval Research Laboratory, where he developed computer applications supporting Research and Development and Environmental Preparedness, Prevention and Response.

Mr. Gallagher holds a Master of Science, Information Systems degree from Syracuse University, a Chief Information Officer Certificate from the National Defense University, and a Bachelor of Science degree in Management Information Systems from James Madison University. He has also completed the Harvard Senior Executive Fellows Program at Harvard University.

Deanna A. Archuleta – U.S. Department of the Interior, Deputy Assistant Secretary for Water and Science

Deanna Archuleta is the Deputy Assistant Secretary for Water and Science in the Department of the Interior. As one of two Deputies in Water and Science, she works with Assistant Secretary Anne Castle to oversee the Bureau of Reclamation and the U.S. Geological Survey.

From January 2008 to April 2009, Archuleta served as board chair of the Bernalillo County (Albuquerque) Water Utility, overseeing the completion of the San Juan Chama Drinking Water Project, one of the largest treatment facilities in the United States. Archuleta also won two terms as a county commissioner in Bernalillo County, and was elected to serve as the chair of the commission in 2009, where she focused on economic development, health care, safety and quality of life initiatives, working to encourage local and national businesses to invest resources in the county, creating local jobs and stimulating the economy.

Before her appointment to the Department, Archuleta was the Southwest Regional Director for the Wilderness Society, engaging with local, State and Federal elected officials as well as a wide variety of stakeholder groups to establish wild land protection throughout the region. She also served on President-Elect Barack Obama's Transition Team in Washington, D.C.

Archuleta received a Master's Degree in Sociology from the University of New Mexico in 2000 and is currently finishing her doctoral degree in Sociology at the university. She received Bachelor of Arts degrees in Sociology and Communications from the University of Washington in Seattle in 1997.

Mark L. DeMulder – Director of the National Geospatial Program, U.S. Geological Survey

Mark DeMulder is the Director of the USGS National Geospatial Program, spanning the bureau's topographic mapping and digital data programs.

Prior to his return to the USGS in 2008, he served as Deputy Director, Office of the Chief Architect, at the National Geospatial-Intelligence Agency (NGA), where he directed enterprise-level architecture and engineering strategies within the National System for Geospatial Intelligence and the NGA corporate enterprise.

Earlier career expertise in geospatial science and national mapping issues include Chief of the Data Policy and Standards Branch in the National Mapping Division of the USGS, where he was the driving force leading the initial design and implementation of *The National Map*. Preceding experience includes Chief of the U.S. Army's Intelligence and Threat Analysis Center's Photogrammetry Branch, and an Imagery Intelligence Officer with the U.S. Air Force.

Mark holds a B.A. degree from the University of Connecticut and an M.S. degree from George Mason University, both in geography. He also is a graduate of the Senior Executive Fellows Program at Harvard University.

He has completed the Department of the Interior's Senior Executive Service Career Development Program. Career highlights include the Open Geospatial Consortium's Vision Award for his work to advance the international geospatial community, the Outstanding Alumni Award from the College of Science at George Mason University, and selection by the U.S. Ambassador to the Permanent Mission of the Organization of American States (OAS) to serve as President of the U.S. National Section to the Pan American Institute of Geography and History, a specialized organization of the OAS.

Frederick Reuss

Frederick Reuss is the author of five novels: *Horace Afoot* (1997), a *New York Times* Notable Book, *Henry of Atlantic City* (1999), awarded the Notable 2000 prize by the American Library Association, *The Wasties* (2002), *Mohr*, and *A Geography of Secrets*. Of the critically acclaimed, *Mohr* (2006), Man Booker Prize winner, John Berger wrote, "His aerialist's sense of history, his sleight of hand, his animal knowledge of political practice, his silver tact and his cool tenderness make his performance nothing less than Orphic." Richard Eder of the New York Times wrote, "Painful and beautiful....Reuss...writes with Jamesian complexity about states of mind and character...with brilliant understanding and a painter's rich detail." His most recent novel, *A Geography of Secrets*, addresses secrecy in public and private life in present-day Washington. It was named a best book of 2010 by the Washington Post.

Dr. Joel D. Scheraga – Senior Advisor for Climate Adaptation, U.S. Environmental Protection Agency

Dr. Joel Scheraga is the Senior Advisor for Climate Adaptation in EPA's Office of Policy in the Office of the Administrator. He is helping EPA integrate considerations of climate change into its programs, regulations, and operations to ensure they are effective under future climatic conditions. He is leading EPA's new Work Group on Climate Change Adaptation Planning, which is charged with developing and implementing a climate change adaptation plan for the Agency. He also represents EPA on the Federal Interagency Climate Change Adaptation Task Force, established by Executive Order in October 2009 to develop recommendations for President Obama on how the Nation might adapt to climate change impacts.

Prior to assuming his current position, Dr. Scheraga served as the National Program Director for EPA's Global Change Research Program in the Office of Research and Development from 1998 to 2009. He participated in the Intergovernmental Panel on Climate Change (IPCC), which was awarded the 2007 Nobel Peace Prize.

Dr. Scheraga is a Fellow of the Institute for Science, Technology and Public Policy in The Bush School of Government and Public Service at Texas A&M University. He is an ex-officio member of the National Research Council Climate Change Education Roundtable. He was lead author of the Human Health chapter of the State of Maryland's *Phase II Strategy for Reducing Maryland's Vulnerability to Climate Change,* released in January 2011.

Dr. Scheraga received an A.B. degree in geology-mathematics/physics from Brown University in 1976, an M.A. in economics from Brown University in 1979, and a Ph.D. in economics from Brown University in 1981.

Larry Sugarbaker – Senior Advisor for *The National Map*, U.S. Geological Survey

Larry Sugarbaker is the National Geospatial Program Senior Advisor for the U.S. Geological Survey. Larry works on *The National Map* policy formulation and new initiatives. He has led major studies to understand customer requirements for *The National Map* and is currently leading a study to assess requirements for a national enhanced elevation program.

Prior to joining the USGS in 2007, Larry was the Vice President and Chief Information Officer for NatureServe, an international non-profit conservation organization. Mr. Sugarbaker worked for the State of Washington, Department of Natural Resources for 22 years, where he managed the geographic information system and supported remote sensing, and forest inventory functions. He has gained worldwide recognition as a leader and expert in geographic information systems. Larry is a past chair of the State of Washington Geographic Information Council and the National Research Council, Mapping Science Committee. Larry graduated from the University of Michigan, School of Natural Resources, with a B.S. in Forestry in 1977. He completed an M.S. degree in Remote Sensing and Wildlife Management from the University of Michigan in 1979. Larry is an active member of the American Society for Photogrammetry and Remote Sensing.

Barbara P. Buttenfield – Professor of Geography,
University of Colorado

Barbara P. Buttenfield is a Professor of Geography at the University of Colorado in Boulder. She teaches courses in Geographic Information Science, Computer Cartography, and Geographic Information Design. Her research interests focus on data delivery on the Internet, visualization tools for environmental modeling, map generalization, and interface usability testing. She has worked extensively with librarians and information scientists to develop Internet-based tools to browse and retrieve information for very large spatial data archives. She led the User Interface Evaluation Team for the first 4 years of the Alexandria Digital Library Project, in collaboration with the University of California - Santa Barbara. She spent several months in residence at the Library of Congress Geography and Map Division while on sabbatical at the U.S. Geological Survey National Mapping Division in Reston, Virginia (1993-1994). She was an original Co-Principal Investigator for the NSF-funded National Center for Geographic Information and Analysis (NCGIA), leading research initiatives on "Multiple Representations," "Formalizing Cartographic Knowledge," and "Visualizing Spatial Data Quality." Dr. Buttenfield is a Past President of the American Cartographic Association, and a Fellow of the American Congress on Surveying and Mapping (ACSM). She was a member of the National Research Council's Mapping Science Committee, 1992-1998. She serves on the Editorial Boards of Annals of the Association of American Geographers; Cartography and GIS; Transactions in GIS; and the URISA Journal.

Acknowledgments

We thank the many scientists whose contributions and accomplishments are reflected in these proceedings, as their efforts ensure continued success for the USGS and *The National Map*. We also acknowledge Stephen S. Aichele, William Carswell, Jacqueline Fahsholtz, Tracy Fuller, Wm. Steve Helterbrand, Kevin Hope, Tim McKinney, Leilani Mock, Mike Mulligan, Daniel Pearson, Robert Pierce, Barbara Ray, Jeff Simley, Ralph Jay Storey, Larry Sugarbaker, Emitt Witt, Shane Wright, E. Lynn Usery, and for their review comments and John Watson and Kay Hedrick for their assistance with the editing and layout of this manuscript. Thanks are extended to the National Training Center staff for their warm hospitality and to Paul Jurasin, William Buster Keaton, and Ryan Stevens for their onsite assistance. Appreciation is also extended to support efforts of the planning committees for organizing a successful USGS GIS Workshop and *The National Map* Users Conference.

U.S. Geological Survey, Geographic Information Science Workshop

Tuesday, May 10, 2011					
ROOM	Earth	Hazards	Polar	Hydro	Ocean	Weeks
	Hands-on Sessions and Seminars			Lecture Sessions		
8:00 AM – 9:45 AM	**The National Map** Introduction to *The National Map* Viewer and Data Delivery Services, Rob Dollison and Matt Tricomi	**Analysis and Modeling** Creating Surfaces from Measurements and Observations, Steve Kopp	**Hydrography Applications** National Hydrography Dataset Applications Workshop, Jeff Simley	**Data Integration: Community for Data Integration** (Introduction by Sky Bristol, no abstract) Fiscal Year 2010 Data Integration Development Project Deliverables—Tools for USGS Scientists, Scott McEwen and Tim Mancuso NWIS Web Services Snapshot Tool for ArcGIS, Sally Holl and others The GeoDataPortal: A Standards-Based Data Access and Manipulation Toolkit for Environmental Modeling, David Blodgett and others	**System Design and Data Managment** EGIS Distribution, Installation, and Licensing of ArcGIS 10, Shane Wright Authoring and Publishing Advanced Geospatial Image Processing Tools to ArcGIS Server, Bob Ternes	**Programming** Developing Tools with ArcGIS Desktop Version 10 Add-Ins, David S. McCulloch Animating MODFLOW Data by Using ArcMap, Steven K. Predmore Migrating Legacy ArcGIS Visual Basic for Applications (VBA) Tools to ArcGIS Version 10 and Visual Basic for .NET, Tana L. Haluska and David S. McCulloch
9:45 AM – 10:15 AM	Break					
10:15 AM – 12:00 PM	**The National Map** Using *The National Map* Services in ArcMap, Matt Tricomi and others	**Analysis and Modeling** New Analysis Capabilities in ArcGIS 10, Steve Kopp	**Hydrography Applications** National Hydrography Dataset Plus (NHDPlus) Version 2 Developments, Tommy Dewald and others	**Data Integration: Community for Data Integration** (Peter Schweizer, no abstract) ScienceBase Data Repository and Catalog—Supporting Geospatial Data Management, Tim Kern and others A Proposal for Using ArcSDE Personal as a Foundation for Developing a USGS Enterprise Geospatial Infrastructure, David S. McCulloch	**Data Managment** Building the Louisiana Seamless GIS Base Map, Sean Deinert and James E. Mitchell Structures and Places Stewardship in Oregon, Bill Clingman Using a Content Management System and GIS for Regional Data Sharing, Matthew Krusemark and others	**Image Processing and Interpretation** Mapping Oil–Water Emulsions from the Deep-water Horizon Oil Spill, G.A. Swayze and others Development of Customized Tools Used in the Creation of the Landsat Image Mosaic of Antarctica (LIMA), Bob Ternes
12:00 PM – 1:30 PM	Lunch					

GIS Workshop

Tuesday, May 10, 2011

ROOM	Earth	Hazards	Polar	Hydro	Ocean	Weeks
	Hands-on Sessions and Seimnars			**Lecture Sessions**		
1:30 PM – 3:15 PM	*The National Map* Web Service Mash-ups with *The National Map* Matt Tricomi and others	**Analysis and Modeling** Using XTools Pro 7 for ArcGIS, Andrei Elobogoev	**Programming** Extending ArcGIS with Python, Drew Flater	**Data Integration** The Coastal and Marine Geoscience Data System; An Open Source Solution to Data Access and Integration, Gregory Miller and Shawn Dadisman Integration of Geology, Geochemistry, and Historical Data within GIS is Helping to Uncover New Mineral Potential in Established Exploration Areas, Birgit Woods	**GIS and Health** GIS Support to CDC's Emergency Operations Center During the Haiti Cholera Response, Brian Kaplan and Michael Wellman Coal Aquifers and Kidney Disease, William Orem and others Open Source News and Citizen Science Tools Aim to Fill the Gaps on Wildlife Health Information, Megan K. Hines and others Geospatial Approaches for Analysis and Modeling of Geographic Distributions of Ticks in the United States, Angela M. James and others	**Online GIS** Deploying a Web Mapping Software Stack: Orthoimagery, Web Server Tuning, WMS, Tile Cache, and Web Map, Dayne Broderson and others Science in Your Watershed: A Graphical Interactive System for Accessing Hydrologic Information in Watersheds of the United States, Michael C. Ierardi
3:15 PM – 3:45 PM	Break					
3:45 PM – 5:30 PM	**Online GIS** Building Great Mashups with ArcGIS.com, Charlie Frye and others	**Mobile GIS** Making Portable Mobile Maps with ArcGIS, Yana Kalinina and others	**Data Management** Best Practices for Image Management and Dissemination with ArcGIS, Cody Benkleman	**Cartographic Design, Output, and Publication: NHD** National Hydrography Dataset (NHD) Generalization Case Study: From Local Resolution New Jersey NHD to 1:24,000 Scale, Ellen L. Finelli and others Metric Assessment of National Hydrography Dataset (NHD) Cartographic Generalization Supporting the 1:24,000-Scale USGS Digital Topographic Map for New Jersey, Lawrence V. Stanislawski and others	**Judicial GIS** Using the Landsat Archive in Court, Melinda McGann Public Access to County GIS Parcel Basemap Data: The Struggle Continues, Bruce Joffe	**Data Management** How will Embedding ArcGIS into SharePoint Will Work in the USGS, Patty Damon and others Consuming SharePoint Lists in ArcMap using SharePoint Web Services, David S. McCulloch
ROOM	Denver Marriott West, Aspen/Snowmass and Beaver Creek/Vail					
7:00 PM – 9:00 PM	**GIS Olympics,** GIS Olympics, Ariel Bates and others					

GIS Workshop
Wednesday, May 11, 2011

ROOM	Earth	Hazards	Polar	Hydro	Ocean	Weeks
	Hands-on Sessions and Seminars			**Lecture Sessions**		
8:00 AM – 9:45 AM	**Cartographic Design, Output, and Publication** Getting Hydrography onto your Maps with ArcGIS 10, Charlie Frye and others	**Image Processing and Interpretation** Working with Landsat Imagery in ENVI: Calibration, Analysis, and Helpful Tips, Amanda O'Connor and Bob Ternes	**Programming** Extending ArcGIS with Python, Drew Flater	**Elevation Data Interpretation** Follow that Stream! Combining LiDAR and the NHD/WBD to Refine North Carolina Watershed Boundaries, Katharine Rainey Kolb and Silvia Terziotti Topographic Effects on Perchlorate Concentrations on a Hillslope at the Amargosa Desert Research Site, Nye County, Nevada, Toby Welborn and others GIS Modeling Techniques to Build an Improved DEM for the Bahamas with SRTM Data, Satellite Imagery, and ALOS PALSAR Data, Sarah M. Trimble	**Analysis and Modeling** Land-cover Mapping Results of Red Rock Canyon National Conservation Area, Clark County, Nevada, J. LaRue Smith Utilizing 3-D Modeling Techniques and Geodatabase Design To Map Placer Gold Occurrences in North Takhar, Afghanistan, Thomas Weeks Moran Development of a GIS Model of Range and Domain of Artisanal Small-Scale Mining Zones of Western Mali, Isabel H. McLoughlin	**Hydrography Applications** StreamStats: An Update on Status, Implementation Process and Future Plans, Kernell Ries and others
9:45 AM – 10:15 AM	Break					
10:15 AM – 12:00 PM	**Cartographic Design, Output, and Publication** Using TerraGo Software for Publishing and Consuming Geospatial PDFs, Michael Bufkin	**Data Integration** Ensuring Accurate, Consistent Measurements with GPS/GNSS, Pamela Fromhertz	**System Design** System Architecture Design, Ty Fabling	**Elevation Data Interpretation** LiDAR: Airborne, Ground, and the Value of Derivative Data, Sean Fitzpatrick The USGS-NGP LiDAR Guidelines and Base Specification, Hans Karl Heidemann	**USGS Science Strategy** Core Science Systems Science Strategy Planning Team: Listening Session for the Eighth Biennial USGS GIS Workshop, Sky Bristol and others	**Hydrography Applications** Colorado Water Science Center Geodatabase of Drainage Basins, Jean A. Dupree and Richard M. Crowfoot Use of the National Hydrography Dataset in StreamStats, Peter A. Steeves Wyoming's Stewardship Efforts to Fix GNIS Issues within the NHD, Paul Caffrey and others
12:00 PM – 1:30 PM	Lunch					

GIS Workshop

Wednesday, May 11, 2011

ROOM	Earth	Hazards	Polar	Hydro	Ocean	Weeks
	Hands-on Sessions and Seminars			**Lecture Sessions**		
1:30 PM – 3:15 PM	*The National Map* *The National Map* Viewer and Services—Train the Trainer, Rob Dollison and others	**Elevation Data Interpretation** LiDAR Use and Management in ArcGIS, Clayton Crawford	**Analysis and Modeling** Building 3-D Subsurface Models in ArcGIS, Doug Gallup and others	**Mobile GIS** From the Desktop to the Field: HTML 5 Mobile Web Mapping Applications, David R. Maltby II High Accuracy Mobile GIS Field Strategy, Alex Mahrou	**GIS and Health** Mineralogical and Geochemical Influences on the 2010 Nigerian Lead Poisoning Outbreak Linked to Artisanal Gold Processing, Geoff Plumlee and others Mapping Oil-Water Emulsions from the Deep-water Horizon Oil Spill, G.A. Swayze and others Spatial Association between Pesticide Use and Levels of Obesity and its Co-morbidities in the United States, Benoit D. Tano	**Data Integration: ADIAS** Building a Framework for Water-Related Data Access: The Ancillary Data Integration and Analysis System (ADIAS), Curtis Price An Overview of National Water Use Information Program Data used by NAWQA Studies, Molly A. Maupin and others Archiving and Maintaining Consistent Watershed Boundaries, Michael E. Wieczorek and others
3:15 PM – 3:45 PM	Break					

GIS Workshop

Wednesday, May 11, 2011

ROOM	Earth	Hazards	Polar	Hydro	Ocean	Weeks
	Hands-on Sessions and Seminars			**Lecture Sessions**		
3:45 PM – 5:30 PM	*The National Map* Intro to Web 2.0 and *The National Map* Direction, Rob Dollison and Matt Tricomi	**Analysis and Modeling** 3-D Feature Analysis in ArcGIS, Clayton Crawford	**Hydrography Applications** NWIS Web Services Snapshot Tool for ArcGIS: A Hands-On Workshop to Retrieve NWIS Data in a Local-Scale Study Area, Sally L. Holl	**Data Management** Demystifying SDE—A Retrospective on the Trials and Tribulations of Implementing ArcSDE as a Master Geospatial Data Library Platform, Nancy A. Damar and Rose L. Medina Enterprise Linear Referencing using the High Resolution National Hydrography Dataset (NHD) and the Hydro Event Management (HEM) Tools: The Oregon Bureau of Land Management (BLM) Experience, Jay Stevens Mapping Out a Course of Action: How the Texas Water Science Center is using Web-Based Applications to Provide and Manage Geospatial Data, and Lessons Learned along the Way, Christy-Ann M. Archuleta	**Analysis and Modeling** Natural-Resources Assessment in Support of Regional Planning and Development in the Anosy Region of Southeastern Madagascar: An Interdisciplinary Geospatial-Based Approach Using Fuzzy Logic, Mark J. Mihalasky A 21st Century Conservation Strategy for America's Great Outdoors, Lisa Duarte and others GIS to the Rescue: Saving the Rio Grande Silvery Minnow, Daniel K. Pearson and J. Bruce Moring	**Data Integration: ADIAS** Creation of SSURGO and Agricultural Practices Datasets at the National Scale, Michael E. Wieczorek GIS Tools for Selection and Area Characterization of Water-Quality Sampling Sites, Curtis Price and Naomi Nakagaki
ROOM	Denver Marriott West, Golden Ballroom					
6:00 PM – 8:00 PM	**Poster Session**					

U.S. Geological Survey, *The National Map* Users Conference

Thursday, May 12, 2011							
8:00 AM – 8:30 AM	Register, Gather, and Network						
8:30 AM – 10:30 AM	Plenary Session – **Kevin Gallagher**, Host Speakers: **Mark DeMulder** – Current Status of *The National Map* and a Vision for the Future **Deanna Archuleta** – Advancing our Geospatial Foundation for Protecting America's Great Outdoors and Powering Our Future **Frederick Reuss** – The Question, "What is a Map?" is More Relevant than Ever						
10:30 AM – 11:00 AM	Break and Networking						
	Panel Discussions						
ROOM	Salon D	Salon E	Salon A B C	Salon F G H	Aspen/Snow	Beaver Creek/ Vail	
11:00 AM – 12:00 PM	**Moderator: Jeff Simley**	**Moderator: Russ Jackson**	**Moderator: Dean Gesch**	**Moderator: Dick Vraga**	**Moderator: Mike Domaratz**	**Moderator: Steve Hammond**	
	Surface Water Mapping Systems— NHD/ NHDPlus/ WBD/NED	**Federal Imagery Partnerships— Status and Plans**	**National Digital Elevation Program—Status and Plans**	**National Transportation Developments**	**Reaching out to *The National Map's* Communities of Users**	**Homeland Infrastructure, Situational Awareness, and Readiness**	
	Speakers: Tommy Dewald, USEPA; Sandra Poppenga, USGS; Seth Hackman, New Jersey; Phillip Henderson, Alabama; Chris Brown, Colorado; Jay Stevens, BLM; Karen Hanson, USGS	**Speakers:** George Lee, USGS; Bill Nellist, NGA; Shirley Hall, USDA; Kent Williams, FSA; NSGIC Representative, FSA	**Speakers:** TBD	**Speakers:** Andrea Johnson, Census Bureau; Steve Coast, Open Street Maps; Marc Levine, USGS; John Gottsegen, TFTN/NSGIC	**Speakers:** Vick Lukas, USGS; Steve Aichele, USGS; Tracy Fuller, USGS	**Speakers:** Talbot Brooks, Delta State; Steve Alness, NGA; Laurie Jasso, USGS; Tammy Barbour, DHS	
						Allie's Grill	Keystone
12:00 PM – 1:30 PM	Lunch/Networking					**Products and Services Feedback** 12:00 PM – 3:30 PM	*The National Map* **Doctor's Office** 12:00 PM – 3:30 PM

The National Map Users Conference

Thursday, May 12, 2011

ROOM	Salon D	Salon E	Salon A B C	Salon F G H	Aspen/Snow	Beaver Creek/Vail	Telluride	Keystone
1:30 PM – 3:00 PM	**NHD 1** WBD/NHD Integration—A New Opportunity for GIS, Stephen Daw Advances in Waterborne Transport Modeling Using NHDPlus, William Samuels Watershed Boundary Dataset (WBD) Applications, Karen Hanson and others	*The National Map* **Data Themes – Elevation** Finding the streams in the Coastal Plain—Will LiDAR help? Silvia Terziotti Status of Alaska Orthoimagery and Elevation Mapping and Alaska Statewide Mapping Program Overview, Tom Heinrichs and Dayne Broderson LiDAR Acquisition of the Atchafalaya River Basin, Louisiana, Jochen Floesser	*The National Map* **Partnerships 1** What Motivates Partnerships—Greed, Altruism or Pragmatism? And Where Do We Go from Here? Jay Parrish New Jersey's Ongoing Partnership with *The National Map*, Seth Hackman and others U.S. Forest Service/U.S. Geological Survey Transportation Data Sharing Initiative, Karen Nabity and others	*The National Map* **Research** Semantic Web Technology for *The National Map*, Dalia Varanka QUAD-G: New Technology for Automated Georeferencing of Scanned Quadrangles, James Burt and others Establishing Classification and Hierarchy in Populated Place Labeling for Multiscale Mapping for *The National Map*, Stephen J. Butzler and others	*The National Map* **and Emergency Operations** The Nationally Consistent Source of Base Mapping, *The National Map* and US Topo: A Sample of The End User's View and Requirements, Maj. William J. Schouviller and Neri G. Terry, Jr. Demonstration: Using *The National Map* as a Common Operating Picture Viewer for Visualization of National Critical Infrastructure and Key Resources, Richard Benjamin *The National Map* and the National Guard: New Synergies for Domestic Operations, Brian Cullis	*The National Map* **Data Themes – Elevation** National Elevation Dataset, Applications of *The National Map*, Sandra Poppenga LiDAR Quality Assurance Challenges and Solutions, Hans Karl Heidemann High Resolution LiDAR in Ecological Research in the Pacific Northwest, Patricia Haggerty	**Board of Geographic Names – Domestic Names Committee** 1:30 PM – 4:00 PM	*The National Map* **Doctor's Office** 12:00 PM – 3:30 PM
3:00 PM – 3:30 PM	**Break and Networking**							

The National Map Users Conference

Thursday, May 12, 2011

ROOM	Salon D	Salon E	Salon A B C	Salon F G H	Aspen/Snow	Beaver Creek/Vail	Telluride	Keystone
3:30 PM – 5:00 PM	**NHD 2** Colorado's Insights to Stewardship of the NHD, Chris Brown National Hydrography Dataset Wet Edit Tool (NHD-WET), Phillip Henderson Applications of the National Hydrography Dataset, Kathy Isham	***The National Map* Data Themes – Historic Maps** Scanning and Georeferencing Historical USGS Quadrangles, Greg Allord ***The National Map* Data Themes – Structures** Geospatial Collaboration Tools Aid Louisiana Structures Project, Craig Johnson and Chris Cretini ***The National Map* Data Themes – Parcels** SEAMLESS USA, Realities and Prospects of a 3,140 County National Parcel Layer, Dennis Klein	***The National Map* Partnerships 2** U.S. Census Bureau and U.S. Geological Survey Partnership for the Decade, Andrea Johnson and Dick Vraga Creating and Utilizing Local-Resolution NHD with LiDAR Data in Florence County, South Carolina, Dave Arnold and others Assessing Impaired Waters Occurrence Within and Near Federal Lands, Tatyana DiMascio and Douglas Norton	***The National Map*-US Topo** The Geospatial PDF: Past, Present, and Future, Michael Bufkin Comparison of Topographic Map Designs for Overlay on Orthoimage Backgrounds, Paulo Raposo and Cynthia Brewer Creating the US Topo—A Process Discussion, Larry Robert Davis	***The National Map* and LiDAR** Accuracy Assessment of a Regional LiDAR DEM, Roy Dokka and others North East LiDAR Project: Project Overview, Michael Shillenn Comparison of Airborne Lidar Elevation Data and USGS National Elevation Dataset Information for Inputs to Regional and Large-Scale Geologic Mapping Applications in Illinois, Donald Luman	***The National Map* Data Themes – Orthoimagery** DigitalGlobe's Advanced Ortho Aerial Program, Kevin Bullock New Jersey US Topo Leaf-off Orthoimage Pilot Project, Roger Barlow and others The Wisconsin Regional Orthophotography Consortium—Building a Statewide Partnership, James Lacy and Dick Vraga	**Board of Geographic Names – Domestic Names Committee** 1:30 PM – 4:00 PM **Legislative Approaches in the Geospatial Community,** Chris Trent 4:00 PM – 5:00 PM	**Listening Session 1**
5:00 PM – 5:30 PM	**Break and Networking**							
5:30 PM – 6:00 PM	**Henry Gannett Award Presentation—Mark DeMulder, host; Deanna Archuleta, Marcia McNutt, Alison Gannett and Family**							
6:00 PM – 8:00 PM	**Post Award Ceremony and Poster Awards—Meet the Author and Plenary Speakers (Refreshments Provided)**							

The National Map Users Conference

Friday, May 13, 2011

ROOM	Salon D	Salon E	Salon A B C	Salon F G H	Aspen/Snow	Beaver Creek/Vail	Telluride	Keystone
8:00 AM – 9:30 AM	Plenary Session – **Mark DeMulder**, Host Speakers: **Dr. Joel Scheraga** – The Importance of Geospatial Information for Effective Adaptation to Climate Change **Larry Sugarbaker** – The National Enhanced Elevation Assessment, Preliminary Findings **Barbara Buttenfield** – Multiple Representations of Geospatial Data: A Cartographic Search for the Holy Grail?							
9:30 AM – 10:00 AM	**Break and Networking**							
10:00 AM – 11:30 AM	**NHD 3** Flow Estimation and the National Hydrography Dataset Plus (NHDPlus) Version 2, Tommy Dewald and others Improving the NHD with Diversion Networks, Kristiana Elite Update on Indiana's Local-Resolution NHD Development and the Geo-Synchronization Web-based NHD Maintenance Project, Phil Worrall and David Nail	**Data Integration** Challenges in Integrating *The National Map* themes of Hydrography and Elevation, Samantha Arundel-Murin Issues in the Concurrent Integration of NHD and NED Updates, Ricardo Lopez-Torrijos and others Canadian-U.S. Hydrographic Data Harmonization and Integration, Michael Laitta	***The National Map* and Other Data 1** Coordination of Activities of the NHD, NHDPlus, WBD, and StreamStats Programs, Kernell Ries and others Integrating Kentucky Karst Data into the National Hydrography Dataset (NHD), James Seay and others Community Maps Program, Charlie Frye (no abstract)	***The National Map* Applications 1** Vegetation Characterization Data made available as WMS Utilizing *The National Map*, M.P. Mulligan and Tim Mancuso *The National Map* Viewer Base Map and Services, Calvin Meyer The National Atlas of the United States 1:1,000,000-Scale Hydrography Dataset: An Overview of the Dataset, the Production Process, and New Mapping Products, Florence Thompson	***The National Map* Applications 2** Management And Population Of Sites: A National Water Information System Application For Automatically Populating Sitefile Information, Steven Predmore and Scott Whitaker An Interactive, GIS-based Application to Estimate Continuous, Unimpacted Daily Streaflow at Ungaged Locations in the Connecticut River Basin, Peter Steeves and Stacey Archfield Use of *The National Map* at FEMA, Doug Bausch and Sara Brush	**Why *The National Map*?** Making the Transition from Paper to Digital Maps, James Mitchell and Sean Deinert *The National Map*: Why Bother? Jay Parrish Standards and Specifications for *The National Map*, Kristin Fishburn	***The National Map* Data Themes – Names** Using the Geographic Names Information System for Interagency Consistency: How the U.S. Census Bureau Has Integrated the USGS Federal Identification Codes into Its Database, Michael Fournier The Changing Role of the Geographic Names Information System: Feature Locations, Corey Plank Collection of Critical Structures and Facilities Data for *The National Map*, Ellen Currier and others	***The National Map* Doctor's Office**
11:30 AM – 12:45 PM	**Lunch/Networking**							

The National Map Users Conference

Friday, May 13. 2011

ROOM	Salon A B C	Aspen/Snow	Salon F G H	Salon D and Salon E	Beaver Creek/ Vail	
12:45 PM – 2:15 PM	*The National Map* **Partnerships 3** From the Ground to the Cloud, from Maps to Web Services: The U.S. Forest Service Experience, Susan DeLost and others Updating Names in the NHD in Oregon: Putting Stewardship to the Test, Robert Harmon and Meredith Carine Geologic Community of Use (CoU), Michael Cooley and others	*The National Map* **and Other Data 2** Integrating Regional Biodiversity Occurrences with *The National Map* Topographic Services—A Test Case for Mobilizing and Visualizing U.S. Geological Survey Science Data Using Distributed Services, Derek Masaki and others Stream View Concept, Alex (Sandy) Williamson Anticipating Plausible Environmental and Related Health Concerns Associated with Future Disasters, Geoff Plumlee and others	*The National Map* **and other Data 3** *The National Map* Science Support Application, Lance Clampitt Mapping Coastal Wetlands in the National Hydrography Dataset—The Louisiana Experience, Sean Deinert and others National Fish Habitat Action Plan Data Delivery, Andrea Ostroff	**Mashathon** Showcase of user applications of *The National Map* data, products, and services.	**Listening Session 2**	
2:15 PM – 3:00 PM	Closing Session: "What You Said: Shaping the Direction of *The National Map*" Adjourn					

Tuesday, May 10, 2011, Geographic Information Science Hands-on Sessions and Seminars

The National Map, 8:00 a.m. – 3:15 p.m.

Introduction to *The National Map* Viewer and Data Delivery Services
by Rob Dollison[1] and Matt Tricomi[2]

[1]U.S. Geological Survey, 12201 Sunrise Valley Drive, MS 511, Reston, VA 20192, (703) 648-5724 rdollison@usgs.gov
[2]Xentity LLC, (720) 244-3063 mtricomi@usgs.gov

This presentation will provide an overview of how new *The National Map* Web services and data download capabilities (http://viewer.nationalmap.gov/viewer/) can be easily accessed and used to support a variety of project needs. Emphasis will be on (1) use of the new *National Map* viewer, (2) learning how to download and access *The National Map* data, and (3) understanding the variety of new Web map services that are available. In addition, the presentation will address the overall future plans and direction of *The National Map* delivery modernization activities for its Web services, basemaps, and data access.

Using *The National Map* Services in ArcMap
by Matt Tricomi,[1] Rob Dollison,[2] Calvin Meyer,[3] and Dan Sandhaus[4]

[1]Xentity LLC, (720) 244-3063 mtricomi@usgs.gov
[2]U.S. Geological Survey, 12201 Sunrise Valley Drive, MS 511, Reston, VA 20192, (703) 648-5724 rdollison@usgs.gov
[3]U.S. Geological Survey, 1400 Independence Road, MS 506, Rolla, MO 65401, (573) 308-3762 clmeyer@usgs.gov
[4]U.S. Geological Survey, 12201 Sunrise Valley Drive, MS 511, Reston, VA 20192, (703) 648-5709 dsandhaus@usgs.gov

This workshop will provide hands-on experience with how to use the new *The National Map* Web services directly in ArcGIS. The presentation will be tailored for ArcGIS desktop users who have little or no experience with *The National Map* viewer or its services and use of geospatial Web services. Attendees will better understand how they can utilize *The National Map* geospatial services in their USGS work.

Web Service Mash-ups with *The National Map*
by Matt Tricomi,[1] Rob Dollison,[2] and Marc Levine[3]

[1]Enterprise Architecture Contractor for National Geospatial Program, Xentity LLC, (720) 244-3063 mtricomi@usgs.gov
[2]U.S. Geological Survey, 12201 Sunrise Valley Drive, MS 511, Reston, VA 20192, (703) 648-5724 rdollison@usgs.gov
[3]U.S. Geological Survey, 12201 Sunrise Valley Drive, MS 511, Reston, VA 20192, (703) 648-6465 mlevine@usgs.gov

Web 2.0 map service concepts such as building mash-ups in various viewers are common and are becoming more standardized and accepted across governments and industry.

This hands-on session will review *The National Map* services and viewer future direction to support these concepts such as:

- How to auto-generate bookmarks from catalogs to open user services in *The National Map* Viewer

- How to save mash-ups with no programming using *The National Map* Viewer for future use or brand as your own instance *The National Map* Viewer

- How to open / save mash-ups in various Web viewers such as Google Maps, Bing Maps, OpenLayers, or various GIS clients such as Google Earth, ArcMap, and ArcGIS Explorer

This workshop will be tailored for more advanced application developers who have little experience with *The National Map* Viewer API (Application Programming Interface) or its services. Attendees will better understand how they can utilize *The National Map* geospatial services and viewer application to create easy mash-ups to provide enhanced user flexibility and productivity by aggregating data and customizing views in the user's viewer or client of choice.

Analysis and Modeling, 8:00 a.m. – 3:15 p.m.

Creating Surfaces from Measurements and Observations
by Steve Kopp

Esri Program Manager, skopp@esri.com

This session will discuss the many kinds of data that GIS users want to create surfaces from and how to choose the correct technique. Some of the example data to be discussed include elevation, water chemistry, population, and some specific concerns such a creating surface from contours or polygons of aggregated statistics. Topics covered will include deterministic and geostatistical interpolation, density surfaces, and other specialized techniques.

New Analysis Capabilities in ArcGIS 10
by Steve Kopp

Esri Program Manager, skopp@esri.com

This session will focus on new analysis capabilities of ArcGIS version 10. Topics to be covered include an overview of Map Algebra and its improvements, new 3-D analysis capabilities, design and modification of environmental sampling networks using statistical techniques, and improvements to ModelBuilder, including iterators.

Using XTools Pro 7 for ArcGIS
by Andrei Elobogoev

Data East, LLC, aelobogoev@dataeast.ru

The 2-hour hands-on session by Data East will familiarize the attendees with XTools Pro, one of the most popular extensions to ArcGIS Desktop. The session will consist of a general overview of the most recent version of the extension, XTools Pro 7, a live demonstration of the product, and hands-on exercises on using the selected, most popular, and powerful tools and features of the extension. Usage tips and tricks will be highlighted as well so that even those participants already using XTools Pro will have a chance to learn something new and become more familiar with all the capabilities of the XTools Pro extension. Any additional discussions, questions, and comments will be welcomed and addressed by the Data East presenters.

Hydrography Applications, 8:00 a.m. – 12:00 p.m.

National Hydrography Dataset Applications Workshop
by Jeff Simley

U.S. Geological Survey, National Geospatial Program, Core Science Systems, Box 25046, Denver Federal Center, MS 510, Denver, CO 80225, (303) 202-4131 jdsimley@usgs.gov

The National Map's National Hydrography Dataset (NHD) is a thorough geospatial representation of the Nation's surface water and can be continually maintained through a community stewardship program. The NHD allows for powerful geospatial analysis using a flow network and linked scientific information. The participation in a workshop can teach many of the fundamentals of the dataset that will allow scientists to exploit the capabilities of the NHD. The NHD was carefully designed for scientific applications and consists of many attributes and characteristics perfectly suited for geographic analysis using GIS. The data were also designed to be simple enough for anyone with a basic GIS skill to use. The workshop provides an overview of NHD applications. The NHD data structure is explored and applied to basic mapping techniques. Basic navigation techniques are covered along with the important process of linking scientific data to the stream network. These techniques, combined with network navigation, lead to cause and effect analysis allowing the scientist to discover how one event in the environment can impact another event. Basic issues of data quality and data maintenance are reviewed.

National Hydrography Dataset Plus (NHDPlus) Version 2 Developments
by Tommy Dewald,[1] Craig Johnston,[2] Rich Moore,[2] Alan Rea,[3] Tim Bondelid,[4] and Cindy McKay[5]

[1]*U.S. Environmental Protection Agency, Office of Water, NHD-NHDPlus Program Manager, dewald.tommy@epa.gov*
[2]*U.S. Geological Survey, New Hampshire - Vermont Water Science Center,311 Commerce Way, Pembroke, NH 03275, (603) 226-7843 cmjohnst@usgs.gov, (603) 226-7825 rmoore@usgs.gov*
[3]*U.S. Geological Survey, Idaho Water Science Center, 230 Collins Rd., Boise, ID 83702, (208) 387-1323 ahrea@usgs.gov*
[4]*Consulting Engineer, Woodville, VA, 540-987-8592 timothy@trbondelid.com*
[5]*Horizon Systems Corporation, Herndon, VA, 703-471-0480 ldm@horizon-systems.com*

The NHDPlus is a suite of geospatial products that builds upon and extend the capabilities of the National Hydrography Dataset (NHD) by integrating it with the National Elevation Dataset (NED) and the Watershed Boundary Dataset (WBD). Interest in estimating NHD streamflow volume and velocity to support pollutant fate-and-transport modeling was the driver behind joint U.S. Environmental Protection Agency (USEPA) and U.S. Geological Survey (USGS) efforts to develop NHDPlus, which was first released in late 2006. NHDPlus has been used in a wide variety of applications since its initial release, and widespread positive responses prompted the multi-agency NHDPlus team to design an enhanced NHDPlus Version 2 that is currently under production and scheduled for release during mid-2011. NHDPlus Version 2 both improves and extends Version 1 data content by leveraging the significantly updated ingredient national datasets. The medium resolution NHD has benefited from thousands of user-supplied updates since 2005. During that time period, an estimated 30 percent of the 30-meter NED has been updated based on resampling of the growing collection of 10-meter elevation data. In addition, where NHDPlus Version 1 used the WBD data for the handful of certified States that were available at the time, Version 2 will include the now complete national coverage for the WBD. The Version 2 data model accommodates the ability to specify the percent of water that travels down each path at major divergences as well as water additions, removals, and inter-basin transfers. Version 2 catchment attributes will again include PRISM temperature and precipitation along with the four 2001 National Land Cover Dataset layers. Over 30,000 USGS streamflow gages, an increase of 7,000 from Version 1, have been located on the NHD network and will be used when producing mean annual and mean monthly streamflow volume and velocity estimates for all networked flowlines in Version 2. These flow estimates will account for the effects of evapotranspiration and are adjusted based upon their network relationships with stream-flow gages in the downstream vicinity. This workshop session will provide an overview of the NHDPlus Version 2 production process and status, including ingredient datasets, data model schema, catchments, international data, attribute tables, data relates, and user tools.

Programming, 1:30 p.m. – 3:15 p.m.

Extending ArcGIS with Python
by Drew Flater

Drew Flater, Esri, Product Engineer, 380 New York Street, Redlands, CA 92373-8100 dflater@esri.com

This session will examine and demonstrate how the Python scripting language can be used to extend ArcGIS analysis and automate mapping. The session is appropriate for those who are somewhat familiar with Python and want to learn about building custom geoprocessing tools, automating workflows, and automating map production.

Online GIS, 3:45 p.m. – 5:30 p.m.

Building Great Mash-ups with ArcGIS.com
by Charlie Frye,[1, 2] Mamata Akella,[1] and Caitlin Scopel[1]

[1]*Esri, 380 New York St., Redlands, CA, 92373-8100, (909) 793-2853*
[2]*Chief Cartographer, cfrye@esri.com*

Great mash-ups start with great data being published in a user-friendly way. However, it is really a bit more than that; those great data must be symbolized to look great against the right basemap. When the combination works, a new map that people want to use and share is the result. This workshop will demonstrate how mash-ups work on ArcGIS.com, what makes a great mash-up, and how to publish data so they are user-friendly, designed for use with a specific basemap, and successfully published so the maps can be used immediately by even the least GIS-savvy users at ArcGIS.com.

Mobile GIS, 3:45 p.m. – 5:30 p.m.

Making Portable Mobile Maps with ArcGIS
by Yana Kalinina, Andrei Elobogoev, and Vyacheslav Ananyev

Data East, LLC, ykalinina@dataeast.ru, aelobogoev@dataeast.ru

This software demonstration by Data East will show how mobile maps can be made using the CarryMap extension to ArcGIS. CarryMap allows for the creation of mobile electronic maps that are an effective replacement for traditional paper maps. These electronic maps are easy to use and have map navigation, identify, search, measure, and GPS functionality included. No software is required for viewing the resulting maps, as they are executable applications themselves (.exe files). As such, they can be easily shared with non-GIS users or used outside of the office environment. Additional points of interest and notes can be added to the map in the field and then uploaded back to ArcGIS in the office. These mobile maps can be used on Windows computers, Windows Mobile handheld devices, iPhones, and iPads. In addition, access to mobile maps can be protected with a password and time limits. A real-world example of this application will be demonstrated.

Data Management, 3:45 p.m. – 5:30 p.m.

Best Practices for Image Management and Dissemination with ArcGIS
by Cody Benkleman

Esri, Technical Product Manager - Imagery, 380 New York Street, Redlands, CA 92373-8100, cbenkelman@esri.com

This session will review recommended methodologies for storing, managing, and serving imagery and metadata, discussing specific processing parameters associated with optical imagery and rasterized elevation data. Different workflows for creating Mosaic Datasets will be reviewed, enabling large collections of imagery to be cataloged and served with dynamic mosaicking and on-the-fly processing. ArcGIS version 10 enables imagery and elevation data to be served as multiple products directly from their source data to a wide range of desktop and Web applications. This session is intended for geospatial professionals who need to work with and make large collections of imagery and elevation data accessible and integrate these data and products into different systems.

Tuesday, May 10, 2011, Geographic Information Science Lecture Sessions

Data Integration: Community for Data Integration, 8:00 a.m. – 12:00 p.m.
Session (8:00 a.m. – 9:45 a.m.) introduction by Sky Bristol (no abstract)

Fiscal Year 2010 Data Integration Development Project Deliverables—Tools for USGS Scientists
by Scott McEwen[1] and Tim Mancuso[2]

[1]U.S. Geological Survey, Box 25046, Denver Federal Center, MS 306, Denver, CO 80225, (303) 202-4663 wsmcewen@usgs.gov
[2]U.S. Geological Survey, Box 25046, Denver Federal Center, MS 302, Denver, CO 80225, (303) 202-4238 tmancuso@usgs.gov

In Fiscal Year 2010, the U.S. Geological Survey (USGS) Community for Data Integration (CDI) sponsored a project to develop a suite of tools and Web services considered to be of high value to USGS scientists who need to find, obtain, and use USGS data for integrated science. The project objectives were divided into three goals: (1) Provide ArcGIS Access to Corporate Databases,

(2) Provide Framework for Loosely Coupling Models, and (3) Create Data Upload, Registry, and Access Tool. The CDI was able to match seed funding with additional funding from partners able to produce the proposed tools and services. Partners include the Texas Water Science Center; National Geospatial Program; Mineral Resources Program; Modeling of Watershed Systems, National Research Program; Center for Integrated Data Analytics; Web Applications Development Team, Fort Collins Science Center; and Biological Informatics Program. The resulting deliverables will be demonstrated in a half-day session. Participant feedback, identification of additional pilot test users, and strategies for future development will be discussed.

NWIS Web Services Snapshot Tool for ArcGIS
by Sally L. Holl,[1] David Maltby,[1] David S. McCulloch,[2] and Brian Reece[1]

[1]*USGS Texas Water Science Center, 1505 Ferguson Lane, Austin, TX 78754, sholl@usgs.gov, drmaltby@usgs.gov, bdreece@usgs.gov*
[2]*U.S. Geological Survey, 12201 Sunrise Valley Drive, MS 445, Reston, VA 20192, (703) 648-5670 dmccullo@usgs.gov*

Retrieving and managing data stored in the U.S. Geological Survey (USGS) National Water Information System (NWIS) database can be a labor-intensive task requiring database training. In 2002 an ArcGIS geodatabase snapshot tool was developed to reduce the steps required to retrieve NWIS data, educate users on the NWIS database, and streamline data management. More than 22 USGS Water Science Centers have utilized this cost-saving snapshot tool to facilitate retrieval and management of NWIS data. In addition, at least four USGS water availability programs have also utilized the snapshot tool (High Plains, Columbia River Plateau, Mississippi Embayment, and the Great Basin). Past snapshot development focused on output directly from an NWIS database installation, typically only available to USGS scientists in Water Science Centers. The 2010 NWIS Web Services Snapshot Tool for ArcGIS represents the next generation of data retrieval and management by enabling instant access to NWIS data from Web services. The new Snapshot still serves as a education tool for new or infrequent NWIS users by translating the numerous codes associated with output files of NWIS data. The Snapshot Tool enables efficient data retrieval and management, which is fundamental to achieving the USGS Science Strategy vision of leveraging its talents and skills to undertake comprehensive and integrated studies that examine the Earth as a system in which the biosphere, hydrosphere, lithosphere, and atmosphere are interrelated.

The GeoData Portal: A Standards-Based Data Access and Manipulation Toolkit for Environmental Modeling
by David L. Blodgett, Tom C. Kunicki, Nate L. Booth, Ivan Suftin, Ryan A. Zoerb, and Jordan I. Walker

U.S. Geological Survey, Center for Integrated Data Analytics, 805 Research Way, Middleton, WI 53562, (608) 821-3899 dblodgett@usgs.gov

Environmental scientists from fields of study including climatology, hydrology, geology, and ecology rely on common data sources and data processing methods. Interest in interdisciplinary science, especially coupling environmental models and data sharing, is increasing among scientists within the U.S. Geological Survey (USGS), other Federal, State, and local agencies, and academia. For example, hydrologic modelers need downscaled climate change projections and land cover data summarized for the watersheds they are modeling to predict streamflow for various climate change scenarios. In turn, ecological modelers are interested in how altered streamflow conditions impact habitat for biotic communities. The USGS Center for Integrated Data Analytics (CIDA) Geo Data Portal (GDP) project formalizes common data acquisition and assimilation tasks to assist in model parameterization, model coupling, and data integration in a standards-based data retrieval and analysis application. Services developed for and adopted by the GDP project simplify and streamline the time consuming and resource intensive tasks that are often barriers to interdisciplinary collaboration. The National Climate Change Wildlife Science Center, Great Lakes Restoration Initiative, and Southeast Regional Assessment Program have provided cases for the development of the GDP. GDP Web services have been designed to leverage and complement standards and software from the geographic, oceanographic, hydrologic, and atmospheric science communities. Interoperability is achieved by adopting existing standards and open-source software components where possible, by working closely with enterprise-scale data providers to serve high value data resources in standard formats, and by developing new transformations and interoperable connections where necessary. *The National Map* has been a major contributor of data for the GDP project by granting access to the National Elevation Dataset and National Land Cover Dataset via robust open-standard service protocols. GDP services allow easy access to raw data from distributed Web servers regardless of data scale or volume. In addition, GDP can dynamically subset, reformat, or summarize accessed data. The GDP project supports simple, rapid development of lightweight user interfaces to commonly needed environmental data access and manipulation tools. Standalone, service-oriented components of the GDP framework provide the metadata cataloging, data subset access, and spatial-statistics calculation needed to support interdisciplinary environmental modeling.

ScienceBase Data Repository and Catalog—Supporting Geospatial Data Management
by Tim Kern,[1] Mike Frame,[2] and Mike Mulligan[3]

[1]*U.S. Geological Survey, Fort Collins Science Center, 2150 Centre Ave., Building C, Fort Collins, CO 80526, (970) 226-9366 kernt@usgs.gov*
[2]*U.S. Geological Survey, Center for Bioinformatics, Core Science Systems, Building 1916T2, 230 Warehouse Road, P. O. Box 6015, Oak Ridge, Tennessee 37831, (865) 576-3605 mike_frame@usgs.gov*
[3]*U.S. Geological Survey, Center for Bioinformatics, Core Science Systems, Box 25046, Denver Federal Center, MS 302, Denver, CO 80225, (303) 202-4242 mmulligan@usgs.gov*

The U.S. Geological Survey (USGS) Community for Data Integration (CDI) worked throughout FY2010 to make data products more accessible to USGS scientists, partner agencies, and the public, sponsoring the Data Integration Development Project to support this enhanced access. As part of this effort, CDI partners developed the ScienceBase Data Repository and Catalog. This application provides a collection point and archive for USGS data products, as well as a discovery and distribution portal for these products. The focus of the ScienceBase Data Repository and Catalog is geospatial data. A number of systems allow users to store and distribute their documents and file-based products. ScienceBase looks to provide much more than just file storage, allowing scientists to upload or register their spatially referenced datasets, and then access a wizard to generate basic metadata. The system supports a variety of data formats, including shapefiles, geotiffs, geodatabases, and model components (inputs, outputs, and code), and automatically generates Web services, maps, charts, graphs, and a number of other data visualizations based on the type of data ingested. The system also allows the user to control access to, and define a release workflow for, these data. This includes support for the USGS Fundamental Science Practices review procedure, encapsulated in a relatively simple Web user interface. True geospatial data management goes well beyond a systems approach. This work would be just one part of the DataONE effort to develop data management tools for principal investigators. The DataONE work attempts to get researchers to consider data management at the inception of their work, and not at the end of the project. The ScienceBase Data Repository and Catalog would make several aspects of geospatial data management less onerous to the scientist and be one of the options that the DataONE effort could provide to users. The presentation includes real-life examples of system use as a data management support tool, focusing on cases involving the USGS Multi-Hazards Project, Great Northern Landscape Conservation Cooperative, and USGS Powell Center.

A Proposal for Using ArcSDE Personal as a Foundation for Developing a USGS Enterprise Geospatial Infrastructure
by David S. McCulloch

U.S. Geological Survey, 12201 Sunrise Valley Drive, MS 445, Reston, VA 20192, dmccullo@usgs.gov

Since the first applications of Geographic Information Systems (GIS) to U.S. Geological Survey (USGS) science, various projects, including most recently the Council for Data Integration, have attempted to develop an enterprise approach to managing geospatial data. These efforts have largely failed for a variety of reasons, leading to an unsatisfactory geospatial data management culture in the USGS. This presentation proposes the development of a USGS enterprise geospatial infrastructure based on Esri's ArcSDE Personal product. ArcSDE Personal is, in turn, based on Microsoft's freely distributable SQL Express Relational Database Management System (RDBMS). The traditional approach is to build databases from the top down with data centers, servers, and centralized management. This presentation proposes the opposite, a bottom-up approach with freely available, enterprise supported client software forming the foundation of an enterprise geospatial infrastructure. The presentation describes how that enterprise geospatial infrastructure can be used with Water Science Center and Regional ArcSDE Enterprise Geodatabases to aggregate geospatial data to the Bureau level.

Data Integration, 1:30 p.m. – 3:15 p.m.

The Coastal and Marine Geoscience Data System; An Open Source Solution to Data Access and Integration
by Gregory Miller and Shawn Dadisman

U.S. Geological Survey, St. Petersburg Coastal and Marine Science Center, 600 Fourth Street South, St. Petersburg, FL 33701, (727) 803-8747 ext. 3020 gmiller@usgs.gov
(727) 803-8747 ext. 3029 sdadisman@usgs.gov

The development of the Coastal and Marine Geoscience Data System (CMGDS, http://cmgds.er.usgs.gov/) is underway and provides online data services to access, manage, and integrate data collected from marine and coastal environments by U.S. Geological Survey (USGS) scientists. Once complete, the CMGDS will provide a single point of access to geoscience data collected by the Coastal and Marine Geology Program (CMGP) using a technology stack of open source Web services and protocols. The CMGDS is comparable to the National Science Foundation-funded Marine Geoscience Data System (MGDS, http://www.marine-geo.org/) and presently serves some CMGP data to their existing 2-D and 3-D earth browsing tools for data integration, visualization, and analysis (GeoMapApp, http://www.geomapapp.org/ and Virtual Ocean, http://www.virtualocean.org/), which are part of the MGDS system). The development of metadata catalogs to complement Geospatial One Stop and Data.gov is also being implemented. Data access is provided using one or more of four methodologies: file download; OGC standard services (WMS, WFS, WCS) and THREDDS (http://www.unidata.ucar.edu/projects/THREDDS/); and via GeoMapApp and Virtual Ocean. Data discovery is provided using an HTML-based interface that provides easy access to data and data publications organized by data type and geographic setting, and metadata catalogs using GeoNetwork and GI-cat. All of the processes are evolving toward providing data access in a manner that will facilitate the integration of CMGP data with other federated datasets such as Data.gov. Population of the database and post-processing of data for use with GeoMapApp and Virtual Ocean has begun. Data were provided for use by these tools because of their widespread use among the marine earth science community, including Federal, State, academic, and international geoinformatics organizations. These tools are also capable of integrating nearly all CMGP data types. The process of integrating CMGP data with the vast resources in GeoMapApp and Virtual Ocean will greatly improve the discovery and use of our data. Serving our data in OGC and other standard protocols will place us in a strong position to easily respond to requests to integrate our data with other Federal systems using standards such as the National Oceanic and Atmospheric Administration's Digital Coast and the U.S. Army Corps of Engineers' eCoastal system.

Integration of Geology, Geochemistry, and Historical Data within GIS is Helping to Uncover New Mineral Potential in Established Exploration Areas
by Birgit Woods

Technical Analyst, Geosoft Inc., birgit.woods@geosoft.com

The more data an exploration geoscientist has, the better. The combination of geophysics, geochemistry, and drilling provides the best results when trying to isolate ore deposits. In this presentation we will look at how the integration of multiple geological data layers such as lake- and stream-sediment samples, geology, topography, and historical drill logs, within a geographic information system (GIS), is helping to define underlying structures, identify outliers and anomalies, and generate new targets in established exploration areas. This presentation will demonstrate the use of mapping and analysis applications that contribute to an integrated workflow for subsurface investigations in GIS, from importing the data, to quality control, analysis, and integration of other datasets, and finally to visualization in a way that is meaningful to colleagues, managers, and industry stakeholders. The presentation will include examples where integrated mapping and analysis techniques are being used to arrive at results that are generating new drill targets in the Nevada goldfields and the Manitoban Flin Flon greenstone belt—both established areas with rich historical data. In Nevada, 3-D mapping, GIS, and a methodical approach to project generation, which combines historical drill logs with different types of data, including ASTER (Advanced Spaceborne Thermal Emission and Reflection Radiometer) data and Digital Elevation Models (DEM) to identify major structures or alternation associated with gold, are helping to generate new prospects for junior explorers. Along the stretch from the Flin Flon–Creighton area on the Manitoba-Saskatchewan border to Lynn Lake, several Canadian geological surveys are working together to conduct targeted mapping and lake-sediment sampling aimed at uncovering new mineral potential. Integrated analysis techniques are being used within a GIS to investigate the geochemistry of existing drill holes in order to piece together a map of the subsurface that—in combination with the geology and geophysics—will serve as a guide to further drilling. The project is funded in part by the Geological Survey of Canada's Targeted Geoscience Initiative (TGI), which is designed to lead to new discoveries in established mining communities. The presentation will also examine how these integrated techniques are making old datasets and knowledge relevant for new projects and incoming geologists.

System Design and Data Management, 8:00 a.m. – 9:45 a.m.

EGIS Distribution, Installation, and Licensing of ArcGIS 10
by Shane Wright

U.S. Geological Survey, Utah Water Science Center, 2329 W Orton Circle, West Valley City, UT 84119,
(801) 908-5067 wright@usgs.gov

Previous to ArcGIS version 10, U.S. Geological Survey (USGS) Enterprise GIS (EGIS) distributed the 200+ gigabyte ArcGIS Suite of software and data to approximately 200 sites by burning multiple copies of the suite to USB hard drives and then shipping them throughout the organization. With improvements in network infrastructure, EGIS, along with USGS Enterprise Active Directory Team (eAD), has developed a Distributed File System (DFS) distribution for ArcGIS 10. The DFS not only saves distribution time and cost but also is much more efficient in keeping the distributed software current with service packs and updates. USGS installation of ArcGIS 10 has been simplified and standardized. The standard installation includes custom tools, features, and fonts. Using the provided installation and uninstall scripts, sites can create a batch file that can then be deployed through the PSEXEC command to automate the installation process. Licensing has been improved and simplified with the release of ArcGIS 10. The popular single-use license is now a standard Esri product and is available to all users under the DOI Enterprise License Agreement (ELA). Concurrent keycodes and dongles have been replaced by provisioning files. Provisioning files are a much more streamlined and efficient method of licensing ArcGIS 10 products.

Authoring and Publishing Advanced Geospatial Image Processing Tools for ArcGIS Server
by Bob Ternes

Senior Account Manager – Federal, ITT Visual Information Solutions, (303) 413-3982 bternes@ittvis.com

Representatives from ITT Visual Information Solutions will demonstrate the development of tools used in an analysis of a time series of satellite imagery to detect changes in natural features. The image processing tools can be authored in ENVI, published to ArcGIS Server, and consumed in ArcGIS Desktop. Using this technique, raster processing experts can create tools that allow ArcGIS users to access sophisticated image processing routines.

Data Management, 10:15 a.m. – 12:00 p.m.

Building the Louisiana Seamless GIS Base Map
by Sean Deinert[1] and James E. Mitchell[2]

[1]*LADOTD/GDM International Services, sdeinert@gdmis.com*
[2]*Louisiana Department of Transportation and Development, IT GIS Manager, jim.mitchell@la.gov*

Maps have been an essential tool in transportation for centuries. To build a road, you need to know where to go and what you will have to traverse. Today, geospatial data and geographic information system (GIS) technology have replaced paper maps. Initially, GIS used the paper map base to digitize and create GIS data. Now, the data make the maps. Despite this paradigm shift, many of today's geospatial data are still derived from inaccurate, out of date maps. Recent technological innovations like digital high-resolution imagery and global positioning systems (GPS) have made data collection faster, more accurate, and cost effective. Combined with GIS, these technologies have transformed map-making from a hand-drawn art form to a digital data-driven, spatial analysis process. In Louisiana, Act 159 of 1928 made what became the Louisiana Department of Transportation and Development (LADOTD) responsible for topographic mapping of the State. Since 2007, DOTD and GDM International Services (GDMIS) have worked to gather the most accurate and best available geospatial data with the goal of creating a seamless, statewide, geospatial database. An important use of this database is to support topographic mapping, as well as a wide range of other applications. In 2010, Governor Jindal signed Act 782 into law, creating Louisiana Revised Statute 48:36, Topographic Mapping. This law places the responsibility of creating a digital geospatial database of Louisiana in the hands of LADOTD. This presentation will discuss the methods used to create a digital map base, innovative approaches developed to address the issues of building and maintaining a digital map base, as well as how a seamless, digital geospatial database is used to produce cartographic products.

Structures and Places Stewardship in Oregon
by Bill Clingman

Senior GIS Analyst, Lane Council of Governments, bclingman@lcog.org

The Oregon Spatial Data Framework encompasses a broad cross-section of geographic data commonly used and shared throughout the State of Oregon. These data elements have been organized into 15 Framework Themes, listed below, each of which is represented by a Framework Implementation Team, or FIT, which coordinates statewide efforts in each area: Administrative Boundaries, Bioscience, Cadastral, Climate, Elevation, Geodetic Control, Geoscience, Hazards, Preparedness, Hydrography, Imagery, Land Use/Land Cover, Transportation, Utilities, and Reference. Some of these 15 Framework Themes include specific data elements that might collectively be referred to as facilities, structures, or cultural landmarks, such as city halls, fire stations, schools, bridges, dams, and so on. While the majority of these facilities/structures data elements fall within the Preparedness theme, many of these individual data elements are of broad general interest, and are utilized in ways that do not fall neatly into a single thematic category, making it difficult to identify a single best source, custodian, or steward for them. At the same time, the immediate needs of specific users must be met. The result is often the evolution of multiple, redundant, and overlapping datasets. This is true at the local and federal level, as well as among State agencies. Hoping to rectify this situation, the U.S. Geological Survey (USGS) entered into a multi-year partnership with the Geospatial Enterprise Office of the Oregon Department of Administrative Services (DAS-GEO) and the Lane Council of Governments (LCOG) to promote and coordinate the development of landmark and structure data within the Oregon Spatial Data Framework and to leverage State efforts in order to improve related federal datasets, specifically the Geographic Names Information System (GNIS) and the National Structures Dataset (NSD). Project goals included the identification of logical data stewards for landmark and structure data in Oregon, as well as the development and adoption of a Stewardship Plan. In pursuit of these goals, a Structures and Places Workgroup was established within the Preparedness FIT. Through the development of a Stewardship Plan for Structures and Places data in Oregon, workgroup participants have identified specific GNIS feature classes and NSD structure types which correspond to recognized Framework data elements and have matched them to existing business practices of various State agencies. These agencies will be the horizontal stewards for one or more specific categories of structure or place feature. The Workgroup is currently developing a Stewardship Charter, as well as a set of detailed maintenance procedures for specific feature types.

Using a Content-Management System and GIS for Regional Data Sharing
by Matthew Krusemark,[1] Jonathan Harahush,[2] and Michael Tafel[3]

[1]Geospatial Team Manager, Denver Regional Council of Governments, mkrusemark@drcog.org
[2]GIS Analyst, DRCOG
[3]GIS Specialist, DRCOG

The Denver Regional Council of Governments (DRCOG) created its first Regional Data Catalog for data sharing in 2008. In 2011, DRCOG decided to integrate an open-source content-management system with the underlying geographic information system (GIS) software foundation present in the current version. Through their data catalog, DRCOG shares data in Open Geospatial Consortium (OGC)-compliant formats with their member (local) governments, the public, and business partners including regional, State, academic, not-for-profit, private sector, and Federal Government. DRCOG would like to share information about the technology foundation, system capabilities, learning curve, and business drivers that led to the latest version of the data catalog being deployed.

Programming, 8:00 a.m. – 9:45 a.m.

Developing Tools with ArcGIS Desktop Version 10 Add-Ins
by David S. McCulloch

U.S. Geological Survey, 12201 Sunrise Valley Drive, MS 445, Reston, VA 20192, dmccullo@usgs.gov

ArcGIS Desktop Version 10 introduces a new development tool called an add-in. Add-ins are based on Microsoft's .NET framework and run on the .NET Common Language Runtime (CLR). This provides numerous advantages over previous ArcGIS Desktop development tools, chief among them being the ability to install custom GIS tools without requiring administrator access or complicated installation procedures. Moreover, add-ins are founded on the idea of a visible Windows control such as a button or a listbox, so for basic tools no user interface code needs to be developed. Add-ins are extremely flexible and powerful, allowing GIS developers to access all ArcObjects, Geoprocessing Toolboxes, and Python scripts. This presentation describes how ArcGIS Desktop Version 10 add-ins are developed, techniques for interacting with other add-in controls, using geoprocessing tools, and executing Python tools.

Animating MODFLOW Data by Using ArcMap
by Steven K. Predmore

U.S. Geological Survey, California Water Science Center, 4165 Spruance Road, Suite 200, San Diego, CA 92101,
(619) 225-6153 spredmor@usgs.gov

Groundwater model simulation results include time series of the spatial distribution of hydrologic heads. Esri has provided tools to animate time-series spatial data in ArcMap. However, these tools are not effective for animating large datasets that have millions of features. For example, the Central Valley hydrologic model has 43,218 spatial features and 512 time-series data for each feature. To animate this dataset with the provided tools, the time-series data must be joined to the spatial features, creating a map layer with over 22 million features. This large numbers of features causes ArcMap to become unresponsive and impractical to use for animations. To overcome these challenges, a Python script was written to help animate the simulated heads. This Python script loops through each set of the time-series data, calculates a field within the spatial layer, renders the layer into ArcMap, and then exports an image of the map before moving to the next time-series data. Once the series of images is created, these images can then be used as frames to make an animation.

Migrating Legacy ArcGIS Visual Basic for Applications (VBA) Tools to ArcGIS Version 10 and Visual Basic for .NET

by Tana L. Haluska[1] and David S. McCulloch[2]

[1]U.S. Geological Survey, 2130 SW 5th Ave., Portland, OR 97216 (503) 251-3212 thaluska@usgs.gov
[2]U.S. Geological Survey, Office of Water Information, MS 445, 12201 Sunrise Valley Dr., Reston, VA 20192, (703) 648-5670 dmccullo@usgs.gov

In ArcGIS versions prior to version 10, Visual Basic for Applications (VBA) was an integral part of the ArcGIS scripting and tool development environment. At version 10 of ArcGIS, VBA is being phased out in favor of Microsoft's .NET application development framework and other development tools, such as those based on the Python computer language. Since subsequent versions of ArcGIS will no longer support VBA, tools written in VBA must be migrated to VB.NET. While similar in syntax, VBA and VB.NET do represent significant migration challenges. This presentation describes the migration experiences and issues encountered in migrating a large ArcMap VBA tool to the new VB.NET development framework. Specific issues such as language syntax differences, development environment changes, and .NET framework considerations are discussed.

Image Processing and Interpretation, 10:15 a.m. – 12:00 p.m.

Mapping Oil-Water Emulsions from the Deepwater Horizon Oil Spill

by G.A. Swayze,[1] R.N. Clark,[1] Ira Leifer,[2] K. Eric Livo,[1] Raymond Kokaly,[1] Todd Hoefen,[1] Sarah Lundeen,[3] Michael Eastwood,[3] Robert O. Green,[3] Neil Pearson,[1] Charles Sarture,[3] Ian McCubbin,[4] Dar Roberts,[5] Eliza Bradley,[5] Denis Steele,[6] Thomas Ryan,[6] Roseanne Dominguez,[7] and the Airborne Visible/Infrared Imaging Spectrometer (AVIRIS) Team

[1]U.S. Geological Survey, Box 25046, Denver Federal Center, MS 964, Denver, CO 80225, (303) 236-0925 gswayze@usgs.gov
[2]Marine Science Institute, University of California, Santa Barbara, CA 93106
[3]California Institute of Technology, Jet Propulsion Laboratory, 4800 Oak Grove Dr., Pasadena, CA 91109-8099
[4]Desert Research Institute, 2215 Raggio Pkwy., Reno, NV 89512
[5]Department of Geography, University of California, Santa Barbara, CA, 93106
[6]National Aeronautics and Space Administration (NASA) Dryden Flight Research Center, P.O. Box 273, Edwards, CA 93523-0273
[7]University Affiliated Research Center, University of California, Santa Cruz/NASA Ames Research Center, Moffett Field, California, CA 94035

The oil spill resulting from the April 20, 2010, Deepwater Horizon disaster provided a chance to test the effectiveness of using UV-NIR imaging spectroscopy for measuring the oil-to-water ratio, sub-pixel areal fraction, thickness, and volume of widespread emulsion slicks (Clark and others, 2010[8]). The NASA/JPL Airborne Visible/Infrared Imaging Spectrometer (AVIRIS) was used to measure the surface reflectance of ocean surface over hundreds of square kilometers where it was covered by thin oil sheens (~0.2 to 6 microns) and thicker water-in-oil emulsions (~1 to 20 millimeters). Measurements were made over the incident site, the surrounding ocean, and coastal areas during four deployments from May to August. Imaging data were corrected to relative reflectance using ground calibration sites located at beaches and airports. Water-in-oil emulsions have a strong UV absorption that imparts concentration-dependent colors in the visible, but are surprisingly bright (up to 60 percent) in reflectance between 1 and 1.3 microns, with diagnostic, in places overlapping, C-H and H_2O absorptions at 0.9, 1.2, 1.4, 1.75, 2, and 2.3 microns. An 80-kilometer boat transect of the spill from the Mississippi Delta out to the incident site provided natural emulsion samples that were reverse-engineered by addition of seawater or evaporation of contained water to form a series of emulsions that spanned nearly the entire oil-to-water ratio range. The diagnostic spectral features of this series, measured over a range of thicknesses, were used to map similar emulsions in AVIRIS data collected during a relatively calm, nearly cloud-free day on May 17, 2010, using the Tetracorder feature-fitting system. This procedure allowed calculation of oil in emulsions on a per-pixel basis giving 19,000 to 34,000 barrels of oil in the AVIRIS scenes. Based on laboratory measurements, near-infrared photons only penetrate a few millimeters into water-in-oil emulsions; thus, the volume of oil derived using this method is a minimum range. Sheens were too thin to exhibit diagnostic vibrational absorptions, and methods for estimating their volume with AVIRIS data are under development. Because of the great extent of the spill, AVIRIS only covered about 30 percent of the core spill area composed of emulsions and sheens. Extrapolation of AVIRIS-derived emulsion density to the core spill area defined on a MODIS (Terra) image collected the same day indicates a minimum of 66,000 to 120,000 barrels of oil floating on the ocean surface that day. This estimate does not include oil in sheens, oil under the surface, oil washed onto beaches and into wetlands, or oil burned, evaporated, or biodegraded as of May 17. Because of the limited penetration of light into emulsions, and based on field observations that emulsions sometimes exceed 20 millimeters in thickness, we estimate that the oil volume, including oil thicker than can be probed with AVIRIS imagery, is possibly as high as 150,000 barrels in the AVIRIS scenes. Extrapolation of this value to the entire spill gives a possible volume of 500,000 barrels for thick oil remaining on the ocean surface as of May 17.

[8]Clark, R.N., Swayze, G.A., Leifer, Ira, Livo, K.E., Lundeen, Sarah, Eastwood, Michael, Green, R.O., Kokaly, Raymond, Hoefen, Todd, Sarture, Charles, McCubbin, Ian, Roberts, Dar, Steele, Doris, Ryan, Thomas, Dominguez, Roseanne, Pearson, Neil, and the Airborne Visible/Infrared Imaging Spectrometer (AVIRIS) Team, 2010, A method for qualitative mapping of thick oil spills using imaging spectroscopy: U.S. Geological Survey Open-File Report 2010–1101, http://pubs.usgs.gov/of/2010/1101/.

Development of Customized Tools Used in the Creation of the Landsat Image Mosaic of Antarctica (LIMA)
by Bob Ternes

Senior Account Manager – Federal, ITT Visual Information Solutions, (303) 413-3982 bternes@ittvis.com

Representatives from ITT Visual Information Solutions will discuss the creation and successful application of custom software tools designed for use in the U.S. Geological Survey (USGS) EROS Data Center. These tools include preprocessing routines, an intuitive desktop interface, and a server-based architecture for final mosaic composition. The tools were used by the USGS to mosaic hundreds of Landsat ETM+ scenes into a continent-wide composite image of Antarctica.

GIS and Health, 1:30 p.m. – 3:15 p.m.

GIS Support to CDC's Emergency Operations Center During the Haiti Cholera Response
by Brian Kaplan[1] and Michael Wellman[2]

[1]Environmental Health Scientist/Geographer, Geospatial Research Analysis and Services Program, Centers for Disease Control and Prevention, (770) 488-3823 bkaplan@cdc.gov
[2]Information Analyst, Geospatial Research Analysis and Services Program, Centers for Disease Control and Prevention, (770) 488-3868 mwellman@cdc.gov

The Geospatial Research, Analysis, and Services Program (GRASP) in the Centers for Disease Control (CDC) has been providing geographic information system support to CDC's Emergency Operation Center (EOC) since 2001. This support includes cartography, visualization, spatial analysis, and system development. GRASP's recent support of CDC's EOC situational awareness activities in response to the Haiti cholera epidemic has pushed GRASP to produce multiple products from the same datasets. These products include maps in pdf format, kml/kmz files for Google Earth, and xml files for Instant Atlas. The automation of these products and goals for the future will be discussed.

Coal Aquifers and Kidney Disease
by William Orem,[1] Calin Tatu,[1] Nikola Pavlovic,[2] Joseph Bunnell,[1] and Robert Finkelman[3]

[1]U.S. Geological Survey, Energy Resources Science Center, Reston, VA, (703) 648-6273; borem@usgs.gov, ctatu@usgs.gov, jbunnell@usgs.gov
[2]Institute of Biomedical Research, Medical Faculty, University of Nis, Nis, Serbia, nikpavster@gmail.com
[3]Department of Geosciences, University of Texas at Dallas, Richardson, TX, (972) 473-7414 bobf@utdallas.edu

Coal, a widely distributed fossil fuel, contains myriad organic and inorganic compounds, some of which are toxic. Human health issues linked to combustion of coal are well known, but little is known about leaching of toxic substances from coal into natural waters, especially drinking water supplies, and impacts on human health. We are investigating the leaching of coal-derived organic substances into drinking water supplies, and related links to human kidney diseases, including cancer. One disease model of impact from coal-derived toxic organic substances in drinking water supplies being studied is Balkan endemic nephropathy (BEN). This kidney disease is restricted to clusters of rural villages in the Balkans. The geographic distribution of BEN coincides with the occurrence of low rank coal. Results from U.S. Geological Survey (USGS) studies suggest that groundwater leaches toxic organic compounds from the coal and transports these compounds to wells/springs used as drinking water sources in the villages. Chronic, long-term (20+ years) exposure to toxic organic compounds in the drinking water appears to be a key factor leading to BEN and cancer of the renal pelvis (RPC). Field studies show that drinking water from BEN villages exhibits higher contents of organic compounds compared to control sites, distal to the coal deposits. Toxicological studies show that organics in water from BEN villages produce excessive cell proliferation in human kidney cells at low dose, and cell death at higher dose. High rates of RPC are also found in parts of the United States with low rank coal deposits and rural populations using coal aquifers as drinking water supplies (for example, in Wyoming, Louisiana, and North and South Dakota). For example, northwestern Louisiana has high rates of RPC and extensive low rank coal deposits (Wilcox Formation). USGS scientists have sampled water from wells completed in the Wilcox coal, and nearby wells outside of the coal-bearing area (control sites). Higher concentrations of potentially toxic organic compounds were observed in water supplies in the coal-bearing area relative to controls and were found to be significantly positively correlated with RPC rates ($p < 0.0001$). The organic compounds present were similar to those observed in well water from BEN areas. Toxicological studies on human kidney cells using dissolved organics isolated from drinking water supplies associated with the Wilcox coal in northwestern Louisiana produced results similar to those from the BEN studies. Ongoing studies are focusing on kidney disease linked to drinking water supplies and the Wilcox coal aquifer in east Texas.

Open Source News and Citizen Science Tools Aim To Fill the Gaps on Wildlife Health Information
by Megan K. Hines,[1] F. Joshua Dein,[2] and C. Marsh[3]

[1]*Technical Manager, Wildlife Disease Information Node, mkhines@wisc.edu*
[2]*U.S. Geological Survey, National Wildlife Health Center, 6006 Schroeder road, Madison, WI 53711, joshua_dein@usgs.gov*
[3]*University of Wisconsin - Madison, Content Manager, Wildlife Disease Information Node, cmarsh@usgs.gov*

The U.S. Geological Survey (USGS) National Biological Information Infrastructure's (NBII) Wildlife Disease Information Node (WDIN), a collaborative project with the USGS National Wildlife Health Center and University of Wisconsin-Nelson Institute of Environmental Studies, works to gather and disseminate information and data on wildlife health topics. These tools enable data sharing, data aggregation, and visualization and promote awareness of wildlife health issues. Two recent applications are highlighted in this session. Launched in 2010, the Wildlife Health Event Reporter (WHER: http://www.wher.org) application contains a series of simple forms allowing anyone (researchers, managers, or the general public) to report sightings of sick or dead wildlife. These reports, when viewed in aggregate, may enable detection of new emerging diseases, biosecurity concerns, or other problems that may affect wildlife. Efforts are under way to market the tool to outdoor enthusiasts and other groups who may be most likely to encounter reportable events. A number of Web services and email alert options are available for offsite notification of event occurrences. Additionally, the WDIN released version 3 of the Global Wildlife Disease News and Information Map application (http://wildlifedisease.nbii.gov/newsmap). The latest enhancements to the toolset bring users increased access to the information behind the map with additional views including a Timeline and Tabular representation of wildlife health events in the news, multiple symbolization options, expanded filtering options, and several Web service choices.

Geospatial Approaches for Analysis and Modeling of Geographic Distributions of Ticks in the United States
by Angela M. James, MaryJane McCool, Philip D. Riggs, MaryAnn Bjornsen, and Jerome E. Freier

USDA, APHIS, VS, CEAH, 2150 Centre Ave, Bldg. B, Fort Collins, CO 80526
Angela.M.James@aphis.usda.gov, MaryJane.McCool@aphis.usda.gov, Philip.D.Riggs@aphis.usda.gov,
MaryAnn.Bjornsen@aphis.usda.gov, Jerome.E.Freier@aphis.usda.gov

Geospatial technologies have had an important impact on the study of vector-borne diseases. We are using geospatial methods to determine the current and potential distribution of arthropod vectors affecting livestock. Because ticks are important vectors of pathogens, knowledge of the geographic distribution of ticks and tick-borne pathogens in the United States is important in developing appropriate targeted surveillance and disease control strategies. A national tick survey program has been developed to monitor tick distributions in the United States. A GIS-based framework is being utilized to integrate two databases into a new geodatabase design. The first includes data from the U.S. National Tick Collection (Smithsonian Institution) and the second is from the U.S. Department of Agriculture's (USDA) National Veterinary Services Laboratories. Both systems contain data regarding distribution, host associations, life cycle, and seasonal activity for a variety of tick species within the United States. The data span as far back as the early 1900s, providing a unique collection of historical information. The newly designed tick geodatabase will be a synthesis of ready-to-map data to produce distribution maps at the county level for tick species of veterinary importance. In addition, these tick location data and a variety of environmental data on climate, vegetation, elevation, and land use will be used to model suitable habitat and predict distributions for specific tick species. We will discuss the framework that we developed for migrating data from the two national databases into a single geodatabase and the methods used to evaluate and map tick distributions. We will also present our current analyses and modeling approaches to explain tick habitat suitability as a measure of the environmental fitness of the habitat for tick survival.

Online GIS, 1:30 p.m. – 3:15 p.m.

Deploying a Web Mapping Software Stack: Orthoimagery, Web Server Tuning, WMS, Tile Cache, and Web Map
by Dayne Broderson, Jay Cable, Tom Heinrichs, and Will Fisher

University of Alaska Fairbanks - Geographic Information Network of Alaska, dayne@alaska.edu, jay@gina.alaska.edu, tom.heinrichs@alaska.edu, will@gina.alaska.edu

Web Mapping Services (WMS) provide a great way to get raster data into the hands of desktop geographic information system (GIS) users and to the general public using Web mapping frameworks like Google Maps. This presentation is an introduction to setting up a Web Mapping Service (WMS) using open source GIS tools and connecting it into tiled Web mapping engines like Google Maps, OpenLayers, and Esri's Viewer for Flex. The entire process of setting up a WMS, including configuring a Web server, preparing raster data, creating the WMS, creating tiles, and finally loading the tiles in a Web mapping framework, will be covered. The software and configuration presented is high-performing and has been developed, tested, and deployed by the University of Alaska Fairbanks (UAF) Geographic Information Network of Alaska in support of the State of Alaska's orthoimagery and DEM portal, AlaskaMapped.org. UAF/Geographic Information Network of Alaska (GINA)/AlaskaMapped serves up over 6.2 terabytes of imagery, elevation, and other datasets via WMS, WCS, and tile interfaces to over 12,000 different users throughout the State of Alaska and the world.

Science in Your Watershed: A Graphical Interactive System for Accessing Hydrologic Information in Watersheds of the United States
by Michael C. Ierardi

U.S. Geological Survey, 12201 Sunrise Valley Drive, MS 445, Reston, VA 20192, (703) 648-5649 mierardi@usgs.gov

The U.S. Geological Survey (USGS) has developed an interactive Web site, Science in Your Watershed, that can be used to access scientific hydrologic and water-resources information organized by 2,264 cataloging unit areas in the United States. Access is provided to the hydrologic cataloging unit area using a user-navigated hierarchical system of mapped watersheds, based on the Water Resources Council hydrologic drainage basin unit (HUC) classification for the United States. This nested drainage basin classification for all 50 States includes 21 water-resources regions, 221 subregions, 378 accounting units, and 2,264 hydrologic cataloging units (that is, HUC_8). Interactive maps of the HUC watersheds include State and county boundaries, major cities, HUC names and HUC code numbers, and stream and river names from the enhanced U.S. Environmental Protection Agency's (USEPA) River Reach File (1:100,000 scale, for example E2RF1). Additional water resource information links are accessible below every 8-digit HUC map accessing the National Water Information System (NWIS) data in the selected HUC watershed. Web links to other hydrologic information are also provided for the selected watersheds, including surface water, groundwater, water quality, National Water-Quality Assessment Program (NAWQA) water-quality data, water-use data, GIS datasets, WaterWatch, nonindigenous aquatic species (NAS) data, and selected USGS water-resources publications. Interagency data links from the National Weather Service Significant River Flood Outlook and the USEPA's Surf Your Watershed, STORET Watershed Stations, and Adopt Your Watershed site can also be accessed for each selected HUC. This collection of water-resource scientific information provides an efficient means to access a comprehensive and consistent set of water-resources information for the United States. Further information can be found at http://water.usgs.gov/wsc.

Cartographic Design, Output, and Publication: NHD, 3:45 p.m. – 5:30 p.m.

National Hydrography Dataset (NHD) Generalization Case Study: From Local Resolution New Jersey NHD to 1:24,000 Scale
by Ellen L. Finelli,[1] Ariel T. Bates,[2] Lawrence V. Stanislawski,[3] and Barbara P. Buttenfield[4]

[1]*U.S. Geological Survey, Applied Research and Technology Branch, Geospatial Technology Section, Box 25046, Denver Federal Center MS 510, Denver, CO 80225, (303) 202-4288 elfinelli@usgs.gov*
[2]*U.S. Geological Survey, TNM Data Operations Branch, Hydrography Section, Box 25046, Denver Federal Center MS 510, Denver, CO 80225, (303) 202-4535 atbates@usgs.gov*
[3]*U.S. Geological Survey, ATA Services, Center of Excellence in Geospatial Information Science (CEGIS), 1400 Independence Road, MS 422, Rolla, MO 65401, (573) 308-3914 lstan@usgs.gov*
[4]*Department of Geography, University of Colorado, Boulder, CO, babs@colorado.edu*

The National Map is a collaborative effort among the U.S. Geological Survey (USGS) and other Federal, State, and local partners to improve and deliver topographic information for the Nation. The partners, through this collaborative effort, produce data at a variety of scales for multiple purposes, thus requiring generalization tools to support consistent 1:24,000-scale (24K) representation. The New Jersey Department of Environmental Protection in partnership with the USGS updated the high-resolution NHD to include 1:2,400-scale hydrographic data. Generalization was performed by the National Geospatial Technical Operations Center (NGTOC), Innovation and Operations offices. The 1:2,400-scale NHD data for New Jersey were generalized to the 24K level of detail using automated generalization tools developed by the USGS Center of Excellence in Geospatial Information Science (CEGIS), with simplification tools developed at the University of Colorado-Boulder. The tools prune the hydrographic network based on the upstream drainage area and target stream density values. State target density values within partitions were calculated from "archived" 24K NHD and applied to achieve the desired partition densities. Generalization methods and results for New Jersey NHD as well as benefits and limitations of the current cartographic generalization procedures will be described. In addition, recommended future enhancements to the NHD generalization procedures will be discussed.

Metric Assessment of National Hydrography Dataset (NHD) Cartographic Generalization Supporting the 1:24,000-Scale USGS Digital Topographic Map for New Jersey
by Lawrence V. Stanislawski,[1] Barbara P. Buttenfield,[2] Ellen L. Finelli,[3] and Ariel T. Bates[4]

[1]*U.S. Geological Survey, ATA Services, Center of Excellence in Geospatial Information Science (CEGIS), 1400 Independence Road, MS 422, Rolla, MO, 65401, (573) 308-3914 lstan@usgs.gov*
[2]*Department of Geography, University of Colorado, Boulder, CO, babs@colorado.edu*
[3]*U.S. Geological Survey, Applied Research and Technology Branch, Geospatial Technology Section, Box 25046, Denver Federal Center MS 510, Denver, CO 80225, (303) 202-4288 elfinelli@usgs.gov*
[4]*U.S. Geological Survey, TNM Data Operations Branch, Hydrography Section, Box 25046, Denver Federal Center MS 510, Denver, CO 80225, (303) 202-4535 atbates@usgs.gov*

The new version of the U.S. Geological Survey (USGS) 1:24,000-scale (24K) topographic map, referred to as US Topo, is compiled from *The National Map* data and can be easily downloaded from the Web in GeoPDF format. Since inception of this new topographic product in 2009, the USGS has been developing automated capabilities to overcome generalization and data integration issues, and subsequently enhance the content of *The National Map* data themes that are included on the digital maps. By overcoming integration obstacles, hydrographic features from the National Hydrography Dataset (NHD) and contours from the National Elevation Dataset (NED) were added to the US Topo production process in October 2009. To further support this modernized mapping effort in the State of New Jersey, the high-resolution NHD was generalized using tools developed by the USGS Center of Excellence for Geospatial Information Science (CEGIS). Collaborative efforts by various levels of government agencies having a variety of needs have produced a multi-scale, high-resolution NHD layer, which was compiled at 1:2,400-scale in New Jersey. Consequently, the 1:2,400-scale New Jersey NHD data were generalized to the 24K level of detail through the CEGIS automated NHD generalization tools. The NHD generalization tools prune the hydrographic network, eliminate extraneous water polygons, and simplify remaining network lines and polygon boundaries. The geoprocessing sequence preserves characteristics important for hydrologic analysis and for cartographic display, such as protecting confluence topology, and retaining overall channel and polygon shapes. In addition, tools have been developed to metrically validate, or assess, resulting generalized datasets and subsequently fine tune generalization processing. The Coefficient of Line Correspondence (CLC) assesses feature content by comparing how well a generalized set of lines match an associated set of lines from a benchmark dataset. Other metrics, such as average segment length, and average area displacement, help evaluate the effects of feature simplification on the geometric characteristics of features. Computation methods for assessment quantities are described in this presentation, along with an assessment of the generalized NHD data for the State of New Jersey.

Judicial GIS, 3:45 p.m. – 5:30 p.m.

Using the Landsat Archive in Court
by Melinda McGann

U.S. Forest Service, Rocky Mountain Region, Remote Sensing Specialist, mlmcgann@fs.fed.us

The U.S. Forest Service used archived Landsat scenes to prove a timber trespass in court. Since then, the U.S. Department of Justice asked for the same support in a multi-jurisdiction water-rights case. This presentation and poster show how we searched the archive, downloaded and processed the scenes, and displayed the images in the courtroom.

Public Access to County GIS Parcel Basemap Data: The Struggle Continues
by Bruce Joffe

Principal, GIS Consultants, also, Founder, Open Data Consortium project, GIS Consultants, 902 Rose Ave., Piedmont, CA 94611, (510) 508-0213 GIS.Consultants@joffes.com

In 2009, the California Court of Appeal affirmed the public's right to Santa Clara County's geographic information system (GIS) basemap data, but the county has raised some new impediments. In 2010, the Sierra Club sued Orange County for access to their GIS parcel basemap, and lost in the trial court. They are now appealing the decision. We will discuss the facts of these cases and their implication for the GIS community.

Data Management, 3:45 p.m. – 5:30 p.m.

How Embedding ArcGIS into SharePoint Will Work in the USGS
by Patty Damon,[1] Jacqueline D. Fahsholtz,[2] Jennifer Sieverling,[3] and Kevin Sigwart[4]

[1]U.S. Geological Survey, Box 25046, Denver Federal Center, MS 801, Denver, CO 80225, (303) 236-4959 pdamon@usgs.gov
[2]U.S. Geological Survey, 230 Collins Road, Boise, ID 83702, (208) 387-1390 jdfahsholtz@usgs.gov
[3]U.S. Geological Survey, Box 25046, Denver Federal Center, MS 801, Denver, CO 80225, (303) 236-1459 jbsiever@usgs.gov
[4]Esri, 8615 Westwood Center Drive, Vienna, VA 22182-2218 703-506-9515 ksigwart@esri.com

ArcGIS Mapping for SharePoint is a set of configurable mapping components that enables SharePoint contributors to embed an interactive map within Microsoft SharePoint. Users of the site can discover and exploit the geographic aspect of their data using an easy-to-understand map. The U.S. Geological Survey (USGS) has implemented SharePoint as an enterprise collaboration tool and widely utilizes ArcGIS; therefore, the combination is intriguing. This session starts with a presentation by Esri and then continues into a panel discussing the possible future of embedding ArcGIS into SharePoint within the USGS environment.

Consuming SharePoint Lists in ArcMap using SharePoint Web Services
by David S. McCulloch

U.S. Geological Survey, 12201 Sunrise Valley Drive, MS 445, Reston, VA 20192, dmccullo@usgs.gov

A SharePoint list is a tabular data format used by SharePoint applications to organize data. Often these SharePoint data have a location or other geospatial component. For example, all USGS offices have a known latitude and longitude, or can be geocoded to a known latitude and longitude. One of SharePoint's strengths is its rich, well documented Web services that expose Share-Point data in a secure way. This presentation demonstrates how geospatial data stored in a SharePoint list can be consumed by

ArcMap using the SharePoint Web services. Specifically it demonstrates how geospatial data stored in the USGS Enterprise SharePoint infrastructure can be consumed by the USGS Enterprise Geographic Information System (GIS) ArcGIS.

GIS Olympics, 7:00 p.m. – 9:00 p.m.

GIS Olympics
by Ariel Bates,[1] Sally Holl,[2] and Daniel Pearson[2]

[1]*U.S. Geological Survey, Box 25046, Denver Federal Center, MS 510, Denver, CO 80225, (303) 202-4535 atbates@usgs.gov*
[2]*U.S. Geological Survey, Texas Water Science Center, 1505 Ferguson Lane, Austin, TX 78754,*
(512) 927-3512 sholl@usgs.gov
(512) 927-3561 dpearson@usgs.gov

To promote awareness of data products made by U.S. Geological Survey geographic science specialists and cooperators, a 'GIS Olympics' event will be held as part of the Geographic Information Science Workshop. Through a series of competitive hands-on challenges, participants will directly interact with data products such as the *The National Map*, National Atlas, National Water Information System, Brazos River Long-Term Monitoring Data Recovery Website, Rio Grande Silvery Minnow Geodatabase, and the Houston Subsidence Map Series. Data products are specifically selected to allow participants to explore and manipulate different geospatial technologies and product formats such as Web sites, Web mapping applications, geospatial software tools, geodatabases, and hard-copy maps. Participants will move between data stations and complete challenge questions over the course of an hour to compete for prizes. Networking and discussion are encouraged before, during, and after the event.

Wednesday, May 11, 2011, Geographic Information Science Hands-on Sessions and Seminars

Cartographic Design, Output, and Publication, 8:00 a.m. – 12:00 p.m.

Getting Hydrography onto your Maps with ArcGIS 10
by Charlie Frye,[1, 2] Mamata Akella,[1] and Caitlin Scopel[1]

[1]*Esri, 380 New York St., Redlands, CA, 92373-8100, (909) 793-2853*
[2]*Chief Cartographer, cfrye@esri.com*

Hydrography is an essential information theme on basemaps and reference maps, yet of all the themes on such maps hydrography is the least well understood and successfully represented. This session will cover some easy ways to leverage the National Hydrography Dataset (NHD) and NHDPlus data. Attendees will learn to symbolize these data in order to make maps ranging from reference maps that are suitable for thematic mapping purposes to more advanced maps like topographic maps or analytical maps. The NHD data are richly attributed and suitable for any of these purposes. Knowing what resources are available makes it easy to include these data in maps.

Using TerraGo Software for Publishing and Consuming Geospatial PDFs
by Michael Bufkin

TerraGo Technologies, Founder and Chief Solutions Architect, mbufkin@terragotech.com, TerraGo Technologies is the producer of a product suite for creation and consumption of Geospatial PDFs.

In this hands-on session, Michael Bufkin, a Founder and Chief Solution Architect for TerraGo, will demonstrate several workflows for creating and using geospatial PDFs. These will include (1) use of Esri ArcMap for GeoSpatial PDF production, (2) round-tripping data from ArcMap to the field and back Using JavaScript to deliver applications with Geospatial PDF, and (3) 3-D GeoPDF production.

Image Processing and Interpretation, 8:00 a.m. – 9:45 a.m.

Working with Landsat Imagery in ENVI: Calibration, Analysis, and Helpful Tips
by Amanda O'Connor and Bob Ternes

ITT Visual Information Solutions, amandao@ittvis.com, bternes@ittvis.com

With 35-plus years of archived Landsat data widely available, a powerful library of archived data can now be harnessed for analysis. Whether you are interested in change detection analysis or land cover classification, Landsat data will provide an accurate solution to obtaining imagery for your projects. Further, if you are interested in fire management, water management, environmental monitoring, or land use planning applications, Landsat data can be used to determine potential fire regions, baseline and historical vegetation mapping, and habitat changes over time. All of the Landsat data can be visualized and analyzed in ENVI with its powerful, workflow-based tools. Join us for this seminar to learn how ENVI can take advantage of the information in Landsat imagery to help you with your analyses. This hands-on seminar will provide an overview and summary of tools for using Landsat data in ENVI and some example applications using Landsat data. Learn how to perform important preprocessing tasks like calibration, do accurate classification and change detection with easy to use workflows, use band ratios to extract detailed information, see tools that integrate ENVI with ArcGIS , and see helpful tips and tricks for multispectral image analysis.

Programming, 8:00 a.m. – 9:45 a.m.

Extending ArcGIS with Python
by Drew Flater

Drew Flater, Esri, Product Engineer, 380 New York Street, Redlands, CA 92373-8100 dflater@esri.com

This session will examine and demonstrate how the Python scripting language can be used to extend ArcGIS analysis and automate mapping. The session is appropriate for those who are somewhat familiar with Python and want to learn about building custom geoprocessing tools, automating workflows, and automating map production.

Data Integration, 10:15 a.m. – 12:00 p.m.

Ensuring Accurate, Consistent Measurements with GPS/GNSS
by Pamela Fromhertz

Colorado State Geodetic Advisor, National Oceanic and Atmospheric Administration, National Geodetic Survey, (303) 236-1468 pamela.fromhertz@noaa.gov

Demands for geospatial accuracy are increasing, and the use of geospatial technologies, such as geographic information systems (GIS), continues to rise. But how often do you have problems with data not aligning in the GIS? Knowing how data are collected in terms of reference systems, coordinate systems, and datums is growing more important to ensure that data layers do align properly. The National Oceanic and Atmospheric Administration's National Geodetic Survey (NGS) provides the basis for the critical geospatial infrastructure called the National Spatial Reference System (NSRS). The NSRS consists of the North American Datum of 1983 (NAD83) and the North American Datum of 1988 (NAVD88). This session will highlight differences such as those between NAD83 and WGS84, why heights on your GPS unit and those on a flood map can be significantly different (for example, 15 meters [50 feet]), and why vertical datums are important to understand, particularly when using Light Detection and Ranging (LiDAR). We will also demonstrate various products that allow you to access the NSRS and discuss planned replacements to both NAD83 and NAVD88 datums.

System Design, 10:15 a.m. – 12:00 p.m.

System Architecture Design
by Ty Fabling

Enterprise Services Architect, 8615 Westwood Center Drive, Vienna, VA 22182-2218, 703-506-9515, tfabling@esri.com

System architecture design provides a solid foundation for deploying and maintaining successful GIS operations. Understanding technology alternatives and selecting appropriate software architecture strategies can mean the difference between implementation success and failure. Hardware and networking technologies support a growing variety of software architecture patterns. Selecting the right technology to support specific business workflow requirements is essential for deployment of successful system operations. This training session will share a system architecture design methodology for supporting successful GIS operations—a methodology developed and proved by successful GIS implementations over the past 18 years. System design strategies are used by Esri customers worldwide to support successful implementation of GIS technology. The session will update users on new developments in capacity planning and performance of Esri software when deployed in today's operational IT environments. The system design models, capacity planning tools, and configuration guidelines provide participants with a proven path to successful GIS operations.

Reference sites: System Design Strategies wiki site
http://wiki.gis.com/wiki/index.php/System_Design_Strategies
System Architecture Design Strategies training class
http://training.esri.com/gateway/index.cfm?fa=catalog.courseDetail&CourseID=50104470_10.X
Esri Press http://esripress.esri.com
Building a GIS book http://esripress.esri.com/display/index.cfm?fuseaction=display&websiteid=141&moduleid=0
Capacity Planning Tool http://esripress.esri.com/display/index.cfm?fuseaction=display&websiteID=141&moduleID=27
Enterprise GIS training blog http://blogs.esri.com/Support/blogs/esritrainingmatters/archive/2011/01/06/
training-spotlight-enterprise-gis.aspx

The National Map, 1:30 p.m. – 5:30 p.m.

The National Map Viewer and Services—Train the Trainer
by Rob Dollison,[1] Matt Tricomi,[2] Calvin Meyer,[3] and Dan Sandhaus[4]

[1]*U.S. Geological Survey, 12201 Sunrise Valley Drive, MS 511, Reston, VA 20192, (703) 648-5724 rdollison@usgs.gov*
[2]*Matt Tricomi, Xentity LLC, (720) 244-3063 mtricomi@usgs.gov*
[3]*U.S. Geological Survey, 1400 Independence Road, MS 506, Rolla, MO 65401, (573) 308-3762 clmeyer@usgs.gov*
[4]*U.S. Geological Survey, 12201 Sunrise Valley Drive, MS 511, Reston, VA 20192, (703) 648-5709 dsandhaus@usgs.gov*

This hands-on workshop is intended for individuals interested in learning how to train *The National Map* users and partners in how to take advantage of *The National Map* Viewer and the new basemap services. An in-depth look at the Viewer, *The National Map* services, and how to download data will all be demonstrated, discussed, and followed-up with exercises to ensure thorough understanding of the topics. This session will be especially useful to U.S. Geological Survey liaisons, and others, interested in helping further the use of *The National Map* data and services through giving training and presentations.

Intro to Web 2.0 and *The National Map* Direction
by Rob Dollison[1] and Matt Tricomi[2]

[1]*U.S. Geological Survey, 12201 Sunrise Valley Drive, MS 511, Reston, VA 20192, (703) 648-5724 rdollison@usgs.gov*
[2]*Xentity LLC, (720) 244-3063 mtricomi@usgs.gov*

This session is an introduction to Web 2.0 concepts and how *The National Map* is enabling wider use of its data and services to support many visualization needs. As the use of map service technology becomes more standardized and accepted across governments and industry, *The National Map* viewer will be able to be used as a portal of authoritative map services which can be integrated—or mashed-up—in many different viewers and GIS Systems, along with other science-based map services or other local datasets. The future direction of *The National Map* data services will be discussed, as well as how these mash-ups, or new user "instances," provide enhanced user flexibility and productivity by aggregating data and customizing views—in the user's viewer of choice—which support specific business needs.

Elevation Data Interpretation, 1:30 p.m. – 3:15 p.m.

LiDAR Use and Management in ArcGIS
by Clayton Crawford

Product Engineer, Esri ccrawford@esri.com
This seminar will outline common LiDAR processing tasks and workflows. Emphasis is on using source point data to produce analytic derivatives. Major workflows include assessing LiDAR point coverage and density for quality control/quality assurance purposes, creating raster digital elevation models (DEMs) and digital surface models (DSMs) from large point collections, estimating forest canopy density and height, minimizing noise for contouring and slope analysis, and delineating floodplains.

Analysis and Modeling, 1:30 p.m. – 5:30 p.m.

Building 3-D Subsurface Models in ArcGIS
by Douglas Gallup[1] (presenter), Thomas Griffiths,[1] Norm Jones,[2] Gil Strassberg,[1] Tim Whiteaker,[3] and Alan Lemon[1]

[1]Aquaveo / Groundwater Consultant dgallup@aquaveo.com, tgriffiths@aquaveo.com, gstrassberg@aquaveo.com, alemon@aquaveo.com
[2]Brigham Young University, njones@byu.edu
[3]University of Texas, twhit@mail.utexas.edu

Building a 3-D subsurface model is a complex task that requires the assembly of many GIS datasets including terrain models, borehole data, geologic maps, faults, and such. These data are combined to derive a set of interpolated datasets that represent the subsurface to our best knowledge. We present a workflow for creating subsurface models within ArcGIS using a combination of standard ArcGIS tools and custom tools that are part of the Subsurface Analyst toolset, developed by Aquaveo in partnership with Esri. The workflow includes classification and visualization of borehole stratigraphy, sketching cross sections within ArcMap, registration of existing cross-section images, interpolation of surfaces, and the generation of 3-D features representing volume models and fence diagrams that can be visualized in ArcScene.

3-D Feature Analysis in ArcGIS
by Clayton Crawford

Product Engineer, Esri ccrawford@esri.com

This seminar emphasizes use of geoprocessing tools that operate on 3-D feature geometry. New capabilities in ArcGIS 10 enable users to do more in true 3-D, not just '2.5-D' surface analysis. Examples include 3-D feature proximity, visibility and skyline analysis with buildings and other above ground features, and volumetric analysis on solids (intersect, difference, union).

Hydrography Applications, 3:45 p.m. – 5:30 p.m.

NWIS Web Services Snapshot Tool for ArcGIS: A Hands-On Workshop to Retrieve NWIS Data in a Local-Scale Study Area
by Sally L. Holl

U.S. Geological Survey Texas Water Science Center, 1505 Ferguson Lane, Austin, TX 78754, (512) 927-3512 sholl@usgs.gov

The 2010 NWIS Web Services Snapshot Tool for ArcGIS represents the next generation of data retrieval and management by enabling rapid access to U.S. Geological Survey (USGS) National Water Information System (NWIS) data. A "Snapshot" is a geodatabase composed of a subset of NWIS data. The Snapshot Tool is geared to address the needs of new or infrequent NWIS users by translating the numerous codes associated with output files of NWIS data into plain English. The Snapshot Tool also enables efficient data retrieval and management, which is fundamental to achieving the USGS Science Strategy vision of undertaking comprehensive and integrated studies that examine the Earth as a system in which the biosphere, hydrosphere, lithosphere, and atmosphere are interrelated. Participants in this hands-on workshop will (1) retrieve NWIS water quality and streamflow data from Web services and compile these data in a personal geodatabase; (2) query water quality data within the geodatabase using Structured Query Language (SQL); (3) import external data such as field study results from an Excel spreadsheet into the geodatabase; and (4) export data for use in analytical software packages such as AquaChem.

Wednesday, May 11, 2011, Geographic Information Science Lecture Sessions

Elevation Data Interpretation, 8:00 a.m. – 12:00 p.m.

Follow that Stream! Combining LiDAR and the NHD/WBD to Refine North Carolina Watershed Boundaries
by Katharine Rainey Kolb[1] and Silvia Terziotti[2]

[1]*U.S. Geological Survey North Carolina Water Science Center, 810 Tyvola Road, Suite 108, Charlotte, NC 28217, (704) 344-6272 X34 kkolb@usgs.gov*
[2]*U.S. Geological Survey North Carolina Water Science Center, 3916 Sunset Ridge Road, Raleigh, NC 27607, (919) 571-4090 seterzio@usgs.gov*

The U.S. Geological Survey North Carolina Water Science Center (USGS NCWSC), in cooperation with the North Carolina Department of Transportation, is expanding the current StreamStats Web-based GIS application, which is implemented only for the Upper French Broad River basin, to cover the entire State of North Carolina. Although the current StreamStats application in western North Carolina uses a local resolution National Hydrography Dataset (NHD), this dataset is not available statewide. Therefore, the statewide StreamStats application will use best available data. North Carolina has a high-resolution elevation dataset derived from Light Detection and Ranging (LiDAR) data, which are used in the NC StreamStats application. As part of the development of the statewide StreamStats application, North Carolina's watershed boundaries will be refined by combining LiDAR-derived surface elevation data and the high resolution (1:24,000 scale) NHD and Watershed Boundary Dataset (WBD). As part of the best available data approach, we sought to create watershed boundaries comparable to the level 4 (8-digit hydrologic unit code, or HUC) national boundaries. In order to find the optimal method for refining the watershed boundaries, four prototype methods of combining the LiDAR and NHD/WBD data were examined. The standard StreamStats method using high-resolution NHD, 10-meter Digital Elevation Models (DEMs), and the WBD served as a control. Three variations were tested as well. The first variation used raw LIDAR-derived catchments with areas of approximately half a square mile to define outer (level 4, 8-digit HUC) watershed boundaries as well as interior drainage boundaries. Since the half-mile catchments are much smaller than the level 6 (12-digit HUC) drainages, the second variation was to group the LiDAR catchments into level 6 (12-digit HUC) "proxies." The final variation was to use LiDAR to define the outer (level 4, 8-digit HUC) boundaries, but not include interior boundaries in the delineation process (neither LiDAR- nor WBD-derived). The best method for creating watershed boundaries comparable to the level 4 (8-digit HUC) national boundaries involved using LiDAR for the outer level 4 (8-digit HUC) boundaries, but leaving out inner drainage boundaries such as 10- and 12- digit hydrologic unit interior boundaries and their LiDAR equivalents.

Topographic Effects on Perchlorate Concentrations on a Hillslope at the Amargosa Desert Research Site, Nye County, Nevada
by Toby Welborn,[1] B.J. Andraski,[1] and W.A. Jackson[2]

[1]*U.S. Geological Survey Nevada Water Science Center, 2730 North Deer Run Road, Carson City, NV 89701,*
(775) 887-7671 twelbor@usgs.gov
(775) 887-7644 andraski@usgs.gov
[2]*Texas Tech University, (806) 742-3449 amdrew.jackson@ttu.edu*

Naturally occurring perchlorate (ClO_4^-) is prevalent in the arid Southwestern United States. The U.S. Geological Survey, in collaboration with Texas Tech University, is investigating the impact of topography on ClO_4^- concentrations to improve the understanding of deposition, accumulation, and biological cycling and transformations in an arid environment. Topography can impact hydrologic, geomorphic, and biologic processes in desert landscapes. At the Amargosa Desert Research Site in southern Nevada, a high-resolution digital terrain model was created to investigate the effects of topographic setting on the magnitude and spatial variability of soil ClO_4^- concentrations across a 9 hectare desert hillslope. Soil samples were collected from shallow (30 centimeters) pits at 50-meter intervals and analyzed for ClO_4^-; distance from each soil collection pit to the nearest plant was measured. Real Time Kinematic GPS data were collected at 0.5-meter intervals and at pit and plant locations using Ashtech Z-Xtreme and Topcon GR-3 dual-frequency receivers. These GPS data were processed to create a 1-meter digital terrain model and derive elevation, slope, aspect, and plan, profile, and mean curvature topographic attributes for the study area. Relationships between ClO_4^- concentrations, topographic attributes, and plant distance were analyzed. Preliminary results indicate a significant inverse correlation between ClO_4^- concentrations and localized elevation. This relation may be suggestive of various processes—hydrologic, geomorphic, and biologic—affecting soil ClO_4^- variability across the hillslope.

GIS Modeling Techniques To Build an Improved DEM for the Bahamas with SRTM Data, Satellite Imagery, and ALOS PALSAR Data
by Sarah M. Trimble

U.S. Geological Survey, Eastern Geology and Paleoclimate Science Center, MS 926-A, 12201 Sunrise Valley Drive, Reston, VA 20192, (703) 648-6061 strimble@usgs.gov

Topographic mapping of tropical areas is typically challenging, due in part to frequent cloud cover, remote locations, or limited ground access. As a result, accurate, up-to-date topographic maps are often unavailable for tropical areas. Global climate change and turbulent weather make accurate topographic maps increasingly necessary—especially in low-lying areas such as the islands of the Bahamas. The solution outlined here creates a detailed topographic model by combining topographic maps, high- resolution satellite imagery, publicly available Digital Elevation Models (DEMs), and Advanced Land Observing Satellite (ALOS) Phased Array type L-band Synthetic Aperture Radar (PALSAR) radar data. The DEM was created using the Topo to Raster (ANUDEM) method designed to develop hydrologically correct DEMs from a variety of input source data. A draft topographic map from the 1950s was the only data available for the island of New Providence. A geographic information system (GIS) was used to digitize the draft's recorded contours and spot heights. Satellite imagery from the IKONOS satellite was examined for ridges and other elevation features that exist between the 50-ft contours; these were digitized as polygons within the GIS. Additional contour data were created by digitizing radar-identified water bodies as polygons, as water bodies maintain a constant elevation throughout. These new polygons were used as additional contours, whose heights were found by averaging the Shuttle Radar Topography Mission (SRTM) elevation value for all pixels falling on the polygon line. Spot height point data from the topographic map were used to define elevation in areas where contours could not be generated. The digitized contours and spot heights from the 1950s draft map, satellite imagery, and PALSAR data were all used as inputs for the Topo to Raster algorithm. The result of these processes is an up-to-date, more accurate topographic model of a tropical, low-relief land mass.

LiDAR: Airborne, Ground, and the Value of Derivative Data
by Sean Fitzpatrick

Director, Strategic Accounts Northeast Sanborn Map Company, 261-17B Langston Ave., Floral Park, NY 11004, (917) 318-9622 sfitzpatrick@sanborn.com, http://www.sanborn.com

LiDAR represents one of the most significant advances in GIS data collection capacity in decades. This presentation will focus on defining LiDAR as an airborne technology and as a ground-based mobile system and will highlight the true value of the derived data in a modern GIS, including an overview of the LiDAR technology itself and its capabilities. Applications that will be discussed include those in transportation, urban development, engineering/construction, vegetation, wetlands mapping, coastal and marine studies, environmental monitoring, change detection, and planimetric mapping. In addition, cost efficiencies and Return on Investment (ROI) will be highlighted. A particular emphasis will be placed on the utility and value of 3-D data with regards to accuracy assessment, methods of managing large datasets, and the advances in Web and cloud technology to enhance the value of LiDAR-derived data. Finally, the presentation will look at future applications of this technology.

The USGS-NGP LiDAR Guidelines and Base Specification
by Hans Karl Heidemann

U.S. Geological Survey, EROS Center, 47914 252nd Street, Sioux Falls, SD 57198, (605) 594-2861 kheidemann@usgs.gov

The 2009 American Recovery and Reinvestment Act stimulus funding brought the U.S. Geological Survey (USGS) National Geospatial Program (NGP) an unprecedented allocation of funds for light detection and ranging (LiDAR) data collection in 2010. With this came a realization that a higher level of consistency for NGP LiDAR data was needed to effectively manage the large volume of point cloud data the NGP would be receiving in 2010-2011. Initially intended to be a short list of basic requirements and simple specifications, the USGS-NGP LiDAR Guidelines and Base Specification (Spec) was born. Widely distributed for comment in 2009-2010, the Spec (drafts v. 12 and v. 13) has been adopted by LiDAR consumers, purchasers, and even governments around the world. With official publication planned this year, data vendors and software developers are building quality assurance tools to assure LiDAR project data and deliverables comply with the requirements of the Spec. The Federal Emergency Management Agency (FEMA) has rewritten its own LiDAR guidelines based on the USGS-NGP document, and the American Society for Photogrammetry and Remote Sensing is developing a set of Best Practice Standards for the LiDAR industry based on many of the guidelines set forth in the document. This session will provide an overview of the USGS-NGP Guidelines and Base Specification highlighting the most significant requirements, explanations of the rationale behind those requirements, and an understanding of how use of the Spec can help unify the LiDAR industry and community and bring us closer to a usable national LiDAR layer.

Analysis and Modeling, 8:00 a.m. – 9:45 a.m.

Land-Cover Mapping Results of Red Rock Canyon National Conservation Area, Clark County, Nevada
by J. LaRue Smith

U.S. Geological Survey, Nevada Water Science Center, 2730 North Deer Run Road, Carson City, NV 89701, (775) 887-7630 jlsmith@usgs.gov

In cooperation with the Bureau of Land Management, the Nevada Water Science Center has produced a land-cover map of the Red Rock Canyon National Conservation Area, Clark County, Nevada using DigitalGlobe's QuickBird high-resolution satellite imagery acquired during the summer of 2006. In collaboration with senior faculty in the Biology Department of the College of Southern Nevada (CSN), Clark County, data at sample field plots were collected in 2008. Data collected included location, environmental descriptions, ground cover, and vegetation descriptions, using the protocol of the U.S. National Vegetation Classification Standard (NVCS). The land-cover map units were delineated at the Alliance level of the NVCS and were merged into categories corresponding to the NVCS hierarchy. Fifty Alliances were merged into 14 Groups, 10 Macrogroups, and 6 Divisions. Other map units include sparse vegetation on outcrops and unconsolidated material, paved road and ruderal communities surrounding the road, disturbed areas, and recently burned areas. Trails were also mapped from the imagery. The workflow to produce the land-cover map included the use of Feature Analyst software, raster processing, and photo interpretation. Sample field plots were used to train and set up learning for Feature Analyst. Results from Feature Analyst were converted to raster to be merged and filtered using raster processing. With the aid of the field biologist from CSN, the results of raster processing were interpreted and edited at the Alliance level of mapping. An accuracy assessment is being conducted. Some of the sample field plots are being used in the accuracy assessment. More general sample field point data called RACE (Rapid Assessment Community Ecology) vegetation samples have been collected from other vegetation mapping programs in southern Nevada. These are being included in the accuracy assessment. It is anticipated that the final land-cover map will be updated from field observations to produce an accurate map to be used for better land management.

Utilizing 3-D Modeling Techniques and Geodatabase Design To Map Placer Gold Occurrences in North Takhar, Afghanistan
by Thomas Weeks Moran

U.S. Geological Survey, 12201 Sunrise Valley Dr., MS 926-A, Reston, VA 20192, (703) 648-6061 tmoran@usgs.gov

This study focuses on the Samti and Nooraba placer gold deposits in the Takhar Province in northern Afghanistan. North Takhar is one of two regions with the largest known placer gold deposits in Afghanistan. While artisanal mining has long been a part of local industry, little has been recorded about the extent of the deposits until recently. The 1980 publication Geology and Mineral Resources of Afghanistan (Abdullah and others, 1980[1]) and U.S. Geological Survey (USGS) Open-File Report 2007-1214, Preliminary non-Fuel Mineral Resource Assessment of Afghanistan, detail all known occurrences of minerals within Afghanistan. Although these reports contain a large amount of valuable information, an advanced estimate of alluvial gold mining capacity using modern remote sensing and GIS techniques has not previously been completed. This paper proposes to convert existing and derived data into a geospatial format in order to more accurately assess if the area could accommodate commercial mining operations and remain economically viable. Additionally, this study will employ three-dimensional (3-D) geospatial modeling techniques to facilitate our understanding of the complex physical and geological relationships of an area's permissiveness for placer gold occurrences. The process of conducting an accurate mineral assessment relies heavily on the availability of accurate data of the area of interest's topography, geomorphology, and geology, each of which influences the origin and displacement of placer deposits. Since transportative processes are directly related to the depositional and formation history of placer deposits, a high-resolution Digital Elevation Model (DEM) is needed for conducting GIS surface modeling. Using a photogrammetric technique known as stereo auto-correlation, a high-resolution DEM (<10 meters) is produced from Advanced Land Observing Satellite (ALOS) Panchromatic Remote-sensing Instrument for Stereo Mapping (PRISM) data. While previous visualization techniques such as overlaying geologic data on top of a DEM are helpful in depicting landscapes, they lack the ability to truly map subsectional data. By superimposing data layers such as overburden depths, gold grade sampling points, and sediment types, spatial queries can be conducted in 3-D, making these variables quantifiable and establishing a basis for surface area and volume calculations. Additionally, through the delineation of mining areas using image interpretation techniques and high-resolution imagery (<1 meter) mining activity was compared with known and permissive placer gold tracts. Through the creation of a well-integrated geodatabase, we performed 3-D spatial and topological queries, and our geospatial analytical capabilities were significantly increased in order to conduct a volumetric and qualitative assessment of placer gold occurrences in northern Takhar.

[1]Abdullah, S.H., Chmyriov, V.M., and Dronov, V.I., 1980, Geology and mineral resources of Afghanistan: Kabul, Afghanistan Geological Survey, 2 volumes.

Development of a GIS Model of Range and Domain of Artisanal Small-Scale Mining Zones of Western Mali
by Isabel H. McLoughlin

U.S. Geological Survey, Geology and Paleoclimate Science Center, 12201 Sunrise Valley Dr., MS 926A, Reston, VA 20192, (703) 640-6047 imclough@mail.umw.edu

This study examines the Kenieba and Bougouni artisanal and small-scale mining (ASM) regions of western Mali. In both of these regions, ASM of gold and (or) diamonds occurs largely as part of the informal economy. ASM, in general terms, refers to the mining activities of small groups or individuals using very basic tools and processes to extract the ore and minerals. These mining practices are often unsustainable and migratory, in that mine sites are abandoned under certain climatic conditions or when yields are low. Oftentimes miners must travel long distances to the mine site and also to other towns to sell their diamonds or gold. This study uses questionnaire/interview information gathered during field work conducted by U.S. Geological Survey (USGS) personnel in February and March of 2007 in Mali. The questionnaires contain information about the mining sites as well as demographic data on the mine workers. From these interviews, information was compiled about the location and size of the mine, the number of workers, gender and tasks of the workers, access routes to the mine, and proximity to local towns or villages. Visualization of these data in a GIS provides a representation of several variables including mine size, number of workers, routes traveled, and nearby towns. Advanced GIS modeling techniques are used to explore how demographic characteristics of the mine workers and spatial data on mine site location and extent can be modeled to provide more information on ASM activities in Kenieba and Bougouni. Specifically, Euclidean distance and Thiessen polygon analysis is compared to Network Analysis using ArcGIS to establish the range and domain of mine worker to mine sites, thereby defining individual areas of influence around each mine site, indicating areas from which mine sites attract workers. The comparison of data from the two ASM regions tests whether there is a correlation between the size of the miner population and the size of the mine, which may then be applied to other countries in West Africa involved in ASM mining practices.

Hydrography Applications, 8:00 a.m. – 12:00 p.m.

StreamStats: An Update on Status, Implementation Process, and Future Plans
by Kernell Ries,[1] John Guthrie,[2] Alan Rea,[3] Martyn Smith,[4] Peter Steeves,[5] and David Stewart[6]

[1] *U.S. Geological Survey, Office of Surface Water, 5522 Research Park Drive, Baltimore, MD 21228, (443) 498-5617 kries@usgs.gov*
[2] *U.S. Geological Survey, Rocky Mountain Mapping Center, Box 25046, Denver Federal Center, MS 516, Denver, CO 80225, (303) 202-4289 jdguthrie@usgs.gov*
[3] *U.S. Geological Survey, Idaho Water Science Center, 230 Collins Road, Boise, ID 83702, (208) 387-1323 ahrea@usgs.gov*
[4] *U.S. Geological Survey, New York Water Science Center, 425 Jordan Road, Troy, NY 12180, (518) 285-5618 marsmith@usgs.gov*
[5] *U.S. Geological Survey, Massachusetts-Rhode Island Water Science Center, 10 Bearfoot Road, Northborough, MA 01532, (508) 490-5054 psteeves@usgs.gov*
[6] *U.S. Geological Survey, Office of Surface Water, 12201 Sunrise Valley Drive, Mail Stop 415, Reston, VA 20192, (703) 648-4879 dwstewar@usgs.gov*

StreamStats (http://streamstats.usgs.gov) is a GIS-based Web application that was developed by the U.S. Geological Survey (USGS) to provide information needed by others for water-resource conservation, planning, and management activities, as well as for engineering design. StreamStats currently is available for use in 24 States. StreamStats allows users to select streamgage locations shown on an interactive map interface to obtain previously published streamflow statistics, basin characteristics, and descriptive information for the stations. Users also can select any ungaged location along a stream to obtain the drainage-basin boundary, basin and climatic characteristics, and estimated streamflow statistics for the location. Estimated streamflow statistics may be obtained by use of regional regression equations or by transferring the flow per unit area at a nearby upstream or downstream streamgage to the ungaged site. StreamStats also includes numerous tools that rely on use of the National Hydrography Dataset (NHD) and the NHDPlus dataset for stream-network navigation to identify and analyze upstream and downstream activities, such as streamgage and dam locations. This session on StreamStats at the workshop will provide an overview of the current application, the implementation process, and plans for the future. The session will include descriptions of the basic tools; stream-network navigation tools; incorporation of water-use summaries for Maryland StreamStats; incorporation of methods for estimating daily mean flows for applications in the Connecticut River Basin in New England, and in Pennsylvania; development of a new, single-user interface for the Nation; use of a batch drainage-basin delineation tool and Web services; the GIS data-preparation process; Water Science Center perspective on data preparation and use of tools to support regional regression studies; and current and potential future configuration of computer resources, and usage statistics.

Colorado Water Science Center Geodatabase of Drainage Basins
by Jean A. Dupree and Richard M. Crowfoot

U.S. Geological Survey, Colorado Water Science Center, Box 25046, Denver Federal Center, MS 415, Denver, CO, 80225, (303) 236-6884 dupree@usgs.gov
(303) 236-6884 crowfoot@usgs.gov

The basin boundary is a fundamental entity used in studies of surface-water resources and for planning water-related projects. Drainage-basin areas published by the U.S. Geological Survey (USGS) in annual Water Data Reports and in the National Water Information System (NWIS) are still primarily based on hardcopy methods. The USGS Colorado Water Science Center developed a geodatabase schema and a basin-delineation method that uses the Watershed Boundary Dataset. This geodatabase includes three feature classes (point, line, and polygon) and four topology rules, which constrain interactions among line and polygon features. Of 217 newly delineated drainage basins, 24 had digital areas that differed by more than 3 percent from published NWIS drainage areas. Nearly all errors were in NWIS drainage-basin areas and were caused by the use of small-scale or poor-quality hardcopy maps. This schema and the delineation method are now used by two other USGS water science centers.

Use of the National Hydrography Dataset in StreamStats
by Peter A. Steeves

U.S. Geological Survey, 10 Bearfoot Road, Northborough, MA 01532, (508) 490-5054 psteeves@usgs.gov

Over the past 2 years, the National StreamStats program (http://water.usgs.gov/osw/streamstats/index.html) has incorporated a suite of tools predicated on the functionality of the National Hydrography Dataset (NHD). Most of these tools showcase the basic navigation functionality of the NHD, and one, the Estimate Flows Based on Similar Streamgaging Stations tool utilizes this basic functionality for hydrologic analysis that would not be possible otherwise. StreamStats and the NHD have proven to be mutually complementary to each other's program objectives. With a focus on advanced Web applications, StreamStats has made NHD available to a broad spectrum of users. Also, through drainage-enforcement pre-processing, StreamStats has opened up the NHD to the National Elevation Dataset (NED) and watershed applications, enhancing the applicability of both NED and NHD in the process. NHD, in turn, has allowed the StreamStats program to delve into vector network navigation functionality. Prior to this, StreamStats was primarily a tool that functioned on top of raster-centric watershed tools. This presentation will step through each of the NHD-centered tools available to users of StreamStats. There will also be discussion on future StreamStats applicability predicted on NHD.

Wyoming's Stewardship Efforts to Fix GNIS Issues within the NHD
by Paul Caffrey,[1] Pat Madsen,[2] and Joel Skalet[3]

[1]GIS Research Scientist/Education Coordinator, University of Wyoming, Wyoming Geographic Information Science Center (WyGISC), Dept. 4008, 1000 E. University Avenue, Laramie, WY 82071, (307) 766-2770 (office) (307) 766-2744 (fax) caffrey@uwyo.edu
[2]GIS Research Scientist, Wyoming Geographic Information Science Center (WyGISC), Dept. 4008, 1000 E. University Avenue, Laramie, WY 82071, (307) 766-2770 patsmad@uwyo.edu
[3]U.S. Geological Survey, NHD Partner Support Cartographer, NGTOC, 505 Science Drive, Madison, WI 53711-1061, (608) 238-9333 ext. 117 jjskalet@usgs.gov

As rich and great as the National Hydrography Dataset (NHD) is, it is not perfect. There will always be a need to improve or enhance the data for the benefit of its users. The Geographic Names Information System (GNIS) data were used to populate the stream names in the NHD as well as many of the hydrologic units included in the Watershed Boundary Dataset (WBD) for the Nation. In general, the process of assigning names and GNIS ID numbers to NHD features was very complete. However, discrepancies persist in every region of the country, in part, due to inherent GNIS line vector coordinate errors as well as interpretative mistakes made during GNIS conflation process when creating the dataset. Under funding provided from the 2009 U.S. Geological Survey (USGS) Partnerships Opportunity Grants, the Wyoming Geographic Information Science Center (WyGISC) has been working with the USGS in an effort to improve and update the National Hydrography Dataset (NHD) and Watershed Boundary Dataset (WBD) for the State of Wyoming. Primary goals for this project are to improve integration of the Geographic Names Information System (GNIS) data with the NHD by updating missing, misplaced, and misspelled GNIS names and the incorrect GNIS feature IDs for the NHD's stream and waterbody features. This presentation will provide others who are now working to improve GNIS named features in the NHD our experience in undertaking the task to correct GNIS errors and to improve GNIS name attribution for major water features within the NHD for the State of Wyoming.

Data Integration: ADIAS, 1:30 p.m. – 5:30 p.m.

Building a Framework for Water-Related Data Access: The Ancillary Data Integration and Analysis System (ADIAS)
by Curtis Price

U.S. Geological Survey, South Dakota Water Science Center, 1608 Mountain View Rd., Rapid City, SD 57702, (605) 394-3242 cprice@usgs.gov

Water-related ancillary data have played a key role in the planning and activities of the U.S. Geological Survey's (USGS) National Water-Quality Assessment (NAWQA) Program from its inception in the early 1990s. These data include both thematic data layers ("coverages") and site-specific ancillary data values computed to characterize individual sampling locations (and related areas such as watersheds or well buffers). Ancillary data have been an integral part of NAWQA planning, data collection, and interpretation. Ancillary data have been critical to interpreting NAWQA water-quality data in a regional and national context. As NAWQA and USGS move forward with their existing water activities, as well as new initiatives, ancillary data will remain important. Identifying and developing ancillary datasets has required substantial effort. There are many different users and producers of water-related ancillary data across the USGS and other Federal agencies, as well as in State governments and non-governmental organizations. Locating and selecting the most suitable datasets ("authoritative data sources") for NAWQA studies has been an important yet difficult task. In the case where needed datasets did not exist, derived datasets have been created, including a set of customized land-use and land-cover datasets for the conterminous United States. Geographic information system (GIS) tools have been developed to address the need to standardize NAWQA ancillary data processing. In addition, data distribution and dataset version-control issues have presented challenges. The Ancillary Data Integration System (ADIAS) is an initiative designed to address these issues. The framework is composed of three main parts: (1) a "portal," which is essentially a Web site that provides the "user's view" of ADIAS; (2) "integrated datasets" for various coverages, aggregated and kept current; and (3) "GIS-derived products and methods" that provide derived coverages and standard tools and techniques for using water-related ancillary datasets. In its initial phase, ADIAS is focused toward meeting NAWQA's ancillary data needs, especially as the Program addresses current planning efforts. ADIAS is being designed to eventually meet the needs of a larger community within USGS, especially in support of other ancillary-data dependent water initiatives such as WaterSMART, and developing linkages with existing data activities such as *The National Map* and the Community for Data Integration.

An Overview of National Water Use Information Program Data used by NAWQA Studies
by Molly A. Maupin,[1] Tammy Ivahnenko,[2] and Marilee Horn[3]

[1]*Molly A. Maupin, U.S. Geological Survey, Idaho Water Science Center, 230 Collins Road, Boise, ID 83702, (208) 387-1307 mamaupin@usgs.gov*
[2]*Tammy Ivahnenko, U.S. Geological Survey, Colorado Water Science Center, Box 25046, Denver Federal Center, MS 415, Denver, CO 80225, (719) 544-7155 ext. 110 ivahnenk@usgs.gov*
[3]*Marilee Horn, U.S. Geological Survey, New Hampshire Water Science Center, 331 Commerce Way, Suite 2, Pembroke, NH 03275, (603) 226-7806 mhorn@usgs.gov*

The U.S. Geological Survey's (USGS) National Water-Quality Assessment (NAWQA) Program uses water-use data for retrospective studies, study planning, and both surface water and groundwater water-quality data analysis. A key source of water-use information for NAWQA has been the National Water Use Information Program (NWUIP). Every 5 years, NWUIP compiles and publishes a national summary report and releases aggregated water-use estimates for counties and hydrologic unit boundaries. NAWQA has successfully used these datasets, but the datasets do not include site-specific information that some studies require. In 2003, NAWQA began efforts to compile and maintain a special National Water Information System (NWIS) database that includes site-specific drinking-water intake and well information within the Site-Specific Water Use Database (SWUDS) system that is part of the NWIS software. This national SWUDS database was constructed from information provided by the U.S. Environmental Protection Agency, Safe Drinking Water Information System (SDWIS); data are linked to site information stored in NWIS. Surface-water intake and groundwater well sites were located, verified, associated with other ancillary data (including links to site information records in NWIS), and entered into an NWIS SWUDS database hosted by the USGS Idaho Water Science Center. This talk will summarize the methods used to construct, maintain, and evaluate the quality of this database.

Archiving and Maintaining Consistent Watershed Boundaries
by Michael E. Wieczorek,[1] Naomi Nakagaki,[2] and James Falcone[3]

[1]U.S. Geological Survey, MD-DE-DC Water Science Center, 5522 Research Park Drive, Baltimore, MD 21228, (443) 498-5550 mewieczo@usgs.gov
[2]U.S. Geological Survey, California Water Science Office, Placer Hall, 6000 J Street Sacramento, CA 95819-6129, (916) 278-3092 nakagaki@usgs.gov
[3]U.S. Geological Survey, 12201 Sunrise Valley Drive, Mail Stop 413, Reston, VA 20192-0002, (703) 648-5008 jfalcone@usgs.gov

Watershed boundaries are essential for the analysis of natural and anthropogenic influences on water quality at sampling sites. The National Water-Quality Assessment (NAWQA) Program has used these watershed boundaries to characterize watersheds at sampling sites by overlaying national-scale thematic map layers representing various landscape characteristics, such as land cover, census variables (including population density), precipitation, soils, and geology in a geographic information system (GIS). When the NAWQA Program began in 1991, GIS specialists accessed archived watershed boundaries and developed others, which were aggregated into a nationwide dataset of basins. Many of these watershed boundaries were delineated by manual methods from 1:100,000- and 1:24,000-scale U.S. Geological Survey (USGS) topographic maps. For large basins, USGS Hydrologic Unit Code (HUC) boundaries were extracted at the 1:250,000 scale. Since then, more detailed watershed boundaries have become available from digital hydrographic data and basin delineation tools, such as Streamstats and NHDPlus. These tools allow the user to develop watershed boundaries at 30-meter resolution. Within the last several years, the USGS, in cooperation with the Natural Resources Conservation Service (NRCS), further divided the HUC system into six levels of hydrologic units at the 1:24,000 scale. This expanded delineation system is known as the Watershed Boundary Dataset (WBD). The availability of detailed watershed boundaries, finer scale hydrographic and elevation data, and GIS tools for basin delineation has greatly facilitated the creation of digital watershed boundaries. The increased availability of watershed datasets and better methods to create watersheds is helpful; however, this has led to a problem in selecting appropriate datasets. For example, multiple watershed boundaries developed for the same site from different data sources and tools result in different watershed boundaries for that site. Some boundaries produced from basin delineation tools are flawed due to limitations that automated methods do not account for. Furthermore, it has become time-consuming to search for the "best" available watershed boundary for a site, especially without a structured archiving system. It is necessary to be able to define the best available watershed dataset when there are several that could be used for watershed characterization. This presentation will go through the history of the different sources of watershed boundaries that have been used by NAWQA and discuss the quality-assurance and quality-control procedures used by NAWQA to evaluate watershed boundaries. Finally, current efforts to aggregate and identify watershed datasets as an "authoritative data sources" for NAWQA as part of the Ancillary Data Integration and Analysis System (ADIAS) will be summarized.

Creation of SSURGO and Agricultural Practices Datasets at the National Scale
by Michael E. Wieczorek

U.S. Geological Survey, MD-DE-DC Water Science Center, 5522 Research Park Drive, Baltimore, MD 21228, (443) 498-5550 mewieczo@usgs.gov

New products have been synthesized by the U.S. Geological Survey at the national scale from existing datasets to enhance investigations related to water resources. Examples of these new datasets include coverages of agricultural tile-drain areas, derived from the 1992 enhanced National Land Cover Data (NLCDe92) land use and the 1992 Natural Resources Conservation Service's (NRCS) National Resource Inventory (NRI); Agricultural practices information such as irrigation, derived from NLCDe92 land use and 1997 NRCS NRI; and soils data, derived from NRCS' Soil Survey Geographic (SSURGO) Database. A major challenge when synthesizing these datasets is spatially apportioning tabular data. The selection of methods to apportion the data is important in the design of datasets so that they can be easily employed by end-users for diverse applications. This talk will review the creation of these layers and their applications. Several projects have successfully used these datasets, including NAWQA(National Water-Quality Assessment) National Synthesis and the National SPARROW (SPAtially Referenced Regressions On Watershed attributes) model, which are used in regional interpretation of water-quality monitoring data that relate in-stream water-quality measurements to spatially referenced characteristics of watersheds.

GIS Tools for Selection and Area Characterization of Water-Quality Sampling Sites
by Curtis Price[1] and Naomi Nakagaki[2]

[1]*U.S. Geological Survey, South Dakota Water Science Center, 1608 Mountain View Rd., Rapid City, SD 57702, (605) 394-3242 cprice@usgs.gov*
[2]*U.S. Geological Survey, California Water Science Center, Placer Hall, 6000 J Street, Sacramento, CA 95819, (916) 278-3092 nakagaki@usgs.gov*

Water-quality assessments typically require development of geospatial datasets for sampling-site selection and statistical characterization of aquifers and watersheds associated with sampling sites. The U.S. Geological Survey (USGS) National Water-Quality Assessment (NAWQA) Program has developed standardized geographic information system (GIS) methods to compute statistics describing landscape characteristics (for example, elevation, land cover, estimated agricultural pesticide application loads, and population density) using spatial overlay analysis, to "thin" point datasets to reduce autocorrelation effects, and to delineate study areas using spatial criteria. These methods were originally implemented in ArcInfo Workstation using the Arc Macro Language (AML), but have recently been rewritten as Python script tools and geoprocessing models in the ArcGIS Desktop environment. A "toolbox" of these scripts compatible with ArcGIS Desktop 9.3 software has been published as "The NAWQA Area-Characterization Toolbox" to allow others to apply these standardized GIS methods to their own water-quality studies. These tools are currently being updated for ArcGIS 10.0. These tools will be maintained in the future as part of activities supported by the Ancillary Data Integration and Analysis System (ADIAS) under development by NAWQA. ADIAS is an initiative to provide water-related ancillary datasets and tools for water studies in NAWQA and elsewhere in USGS. The NAWQA Area-Characterization Toolbox will be used to update ancillary data computations required as new surface-water and groundwater sites and studies are added to the NAWQA Program.

USGS Science Strategy, 10:15 a.m. – 12:00 p.m.

Core Science Systems Science Strategy Planning Team: Listening Session for the Eighth Biennial USGS GIS Workshop
by Sky Bristol, Chip Euliss, Barbara Poore, Brian McCallum, David Miller, Dean Gesch, Jay Diffendorfer, Nate Booth, Nina Burkardt, Rich Signell, Roland Viger, and Suzette A. Morman

Core Science Systems Science Strategy Planning Team: sbristol@usgs.gov, ceuliss@usgs.gov, bspoore@usgs.gov, bemccall@usgs.gov, dmiller@usgs.gov, gesch@usgs.gov, jediffendorfer@usgs.gov, nlbooth@usgs.gov, burkardtn@usgs.gov, rsignell@usgs.gov, rviger@usgs.gov, smorman@usgs.gov

The Core Science Systems (CSS) Science Strategy Planning Team will hold a listening session at the 2011 GIS and National Map workshop to discuss the goals and major themes of the developing document and receive input toward the final product. The goal of the team is for as many U.S. Geological Survey (USGS) staff and partners as possible to see their input clearly articulated in the final report that will set the course for Core Science Systems in the coming decade. The team looks forward to hearing about bright spots in the organization where amazing work in data integration and scientific information systems is being conducted. The team also wants to hear ideas on how the USGS can best go on providing a more robust scientific data platform for research and technical assistance to decision makers. The strategic science planning process began with the USGS Science Strategy in 2007 and continues with a group of planning teams working on the seven mission areas that have emerged to address societal issues. Core Science Systems embodies the major USGS Programs brought together into the new Mission Area under the 2011 realignment along with strong partnerships throughout USGS Science Centers. Development of programs such as *The National Map* and the National Cooperative Geologic Mapping Program through the information science and data integration initiatives suggested in the Science Strategy will provide the geospatial fabric upon which USGS scientists can envision and conduct projects of all kinds and deliver actionable, integrated information to the world. Through the computer, information, and library science aspects of the organization and an organizational commitment to enabling and facilitating the USGS-wide community of researchers, data managers, and other specialists, Core Science Systems is poised to continually provide the best available scientific computing platform as the bedrock for USGS science.

Mobile GIS, 1:30 p.m. – 3:15 p.m.

From the Desktop to the Field: HTML 5 Mobile Web Mapping Applications
by David R. Maltby II

*U.S. Geological Survey, Texas Water Science Center,1505 Ferguson Lane, Austin, TX 78754,
(512) 927-3588 drmaltby@usgs.gov*

Mobile devices such as the iPhone, Android, and iPad with HTML 5 are providing new and inexpensive ways to take mapping applications into the field for data collection, emergency management, and navigational routing. Web-standard interfaces allow individuals to send and retrieve data through the Internet, enabling the seamless integration of content-rich, well- formed data into smart phones and tablet device applications. Building a Web-map and data interface takes little time, but the results can be eye-catching and productive. This presentation will demonstrate mobile mapping applications for scientists and natural resource managers using location-aware Google Maps, ArcGIS Server, and Web services on mobile devices. An introduction to developing mobile Web applications using HTML 5 is also included.

High Accuracy Mobile GIS Field Strategy
by Alex Mahrou

RockyMountainGeo-GIS Consultant, alex@rockymountaingeo.com

Accurate geographic information system (GIS) data are needed more than ever in utility, oil and gas, and municipal operations and in environmental studies. As the use of GIS becomes more ubiquitous across new realms, questions will be asked regarding the spatial integrity of your data. How can you ensure that the recipient of your data knows that they are accurate? This Mobile GIS seminar will demonstrate how to prove the horizontal and vertical accuracy of your GIS data by using professional-grade GPS hardware, National Geodetic Survey (NGS) Monuments, NGS Datasheet know-how, and a few other rudimentary tools. Attendees will walk away from this session with the utmost confidence that the acronym GIS does not stand for "Get It Surveyed."

GIS and Health, 1:30 p.m. – 3:15 p.m.

Mineralogical and Geochemical Influences on the 2010 Nigerian Lead Poisoning Outbreak Linked to Artisanal Gold Processing
by Geoff Plumlee,[1] James Durant,[2] Antonio Neri,[3] Suzette Morman,[4] Ruth Wolf,[5] Greg Meeker,[6] Carrie Dooyema[7]

[1]*U.S. Geological Survey, Box 25046, Denver Federal Center, MS 973, Denver, CO 80225, gplumlee@usgs.gov*
[2]*CDC/ATSDR, Emergency Response Coordinator, hzd3@cdc.gov*
[3]*M.D., MPH, CDC, bro0@cdc.gov*
[4]*U.S. Geological Survey, Box 25046, Denver Federal Center, MS 964, Denver, CO 80225, smorman@usgs.gov*
[5]*U.S. Geological Survey, Box 25046, Denver Federal Center, MS 964, Denver, CO 80225, rwolf@usgs.gov*
[6]*U.S. Geological Survey, Box 25046, Denver Federal Center, MS 973, Denver, CO 80225, gmeeker@usg.sgov*
[7]*CDC, Epidemic Intelligence Service Officer, igb7@cdc.gov*

In spring 2010, a pattern of ongoing childhood deaths (perhaps as high as 400) and illness (355+ cases) was noted by Medecines Sans Frontieres (MSF) in several villages of Zamfara State, Nigeria, and was determined to be due to lead poisoning. The presumed cause was artisanal gold ore processing, which had recently become more mechanized. In the villages, gold ores are first hand crushed, then pulverized in a flourmill (the recent mechanization), then washed and the gold extracted by amalgamation (liquid mercury intermixed with the ore by hand, then the mixture heated by blowtorch). In May 2010, the U.S. Centers for Disease Control (CDC) deployed a team to two of the hardest-hit villages to help assess the extent of the lead poisoning, characterize the sources of exposure to lead and other heavy metals, and recommend measures to mitigate exposures. The CDC team found unprecedented levels of acute lead poisoning, particularly in young children, with blood lead levels as high as 400 micrograms per deciliter (>10 is indicative of lead poisoning). The CDC Agency for Toxic Substances and Disease Registry (ATSDR) team collected an extensive sample suite, including raw and ground ores, soils, waters, and dust sweep samples. The U.S. Geological Survey (USGS) is working with CDC to determine mineralogical and geochemical characteristics of the solid samples, to aid in exposure assessment. Mineralogical analyses indicate that ores being processed are lightly to moderately oxidized quartz vein ores. Primary lead sulfide (which is not bioaccessible) is present in levels ranging from 1 to ~30 volume percent in unoxidized ore samples. Weathering and oxidation prior to mining created a complex secondary lead mineral assemblage dominated by highly bioaccessible lead carbonate, with lesser lead phosphate, oxide, sulfate, vanadinate, and tungstate minerals. The same lead minerals are abundant in dust sweep samples collected from eating areas of many family compounds where the ores were processed, indicating severe contamination resulted from the processing and a logical exposure route via hand/mouth transmission and incidental ingestion. Bulk lead concentrations in the sweep samples can be as high as 12 percent (> 0.04 percent lead in soil is considered elevated), much of which is indicated to be bioaccessible by physiologically based extraction tests using simulated gastric fluids. Bulk mercury is also high in the sweep samples (up to 30 parts per million), indicative of contamination from mercury lost during the amalgamation; some of this mercury is water-soluble and bioaccessible, indicating that mercury contamination may also be problematic from an exposure and possible toxicological perspective.

Mapping Oil-Water Emulsions from the Deepwater Horizon Oil Spill

by G.A. Swayze,[1] R.N. Clark,[1] Ira Leifer,[2] K. Eric Livo,[1] Raymond Kokaly,[1] Todd Hoefen,[1] Sarah Lundeen,[3] Michael Eastwood,[3] Robert O. Green,[3] Neil Pearson,[1] Charles Sarture,[3] Ian McCubbin,[4] Dar Roberts,[5] Eliza Bradley,[5] Denis Steele,[6] Thomas Ryan,[6] Roseanne Dominguez,[7] and the Airborne Visible/Infrared Imaging Spectrometer (AVIRIS) Team

[1]*U.S. Geological Survey, Box 25046, Denver Federal Center, MS 964, Denver, CO 80225, (303) 236-0925 gswayze@usgs.gov*
[2]*Marine Science Institute, University of California, Santa Barbara, CA 93106*
[3]*California Institute of Technology, Jet Propulsion Laboratory, 4800 Oak Grove Dr., Pasadena, CA 91109-8099*
[4]*Desert Research Institute, 2215 Raggio Pkwy., Reno, NV 89512*
[5]*Department of Geography, University of California, Santa Barbara, CA, 93106*
[6]*National Aeronautics and Space Administration (NASA) Dryden Flight Research Center,*
P.O. Box 273, Edwards, CA 93523-0273
[7]*University Affiliated Research Center, University of California, Santa Cruz/NASA Ames Research Center,*
Moffett Field, California, CA 94035

The oil spill resulting from the April 20, 2010, Deepwater Horizon disaster provided a chance to test the effectiveness of using UV-NIR imaging spectroscopy for measuring the oil-to-water ratio, sub-pixel areal fraction, thickness, and volume of widespread emulsion slicks (Clark and others, 2010[8]). The NASA/JPL Airborne Visible/Infrared Imaging Spectrometer (AVIRIS) was used to measure the surface reflectance of ocean surface over hundreds of square kilometers where it was covered by thin oil sheens (~0.2 to 6 microns) and thicker water-in-oil emulsions (~1 to 20 millimeters). Measurements were made over the incident site, the surrounding ocean, and coastal areas during four deployments from May to August. Imaging data were corrected to relative reflectance using ground calibration sites located at beaches and airports. Water-in-oil emulsions have a strong UV absorption that imparts concentration-dependent colors in the visible, but are surprisingly bright (up to 60 percent) in reflectance between 1 and 1.3 microns, with diagnostic, in places overlapping, C-H and H_2O absorptions at 0.9, 1.2, 1.4, 1.75, 2, and 2.3 microns. An 80-kilometer boat transect of the spill from the Mississippi Delta out to the incident site provided natural emulsion samples that were reverse-engineered by addition of seawater or evaporation of contained water to form a series of emulsions that spanned nearly the entire oil-to-water ratio range. The diagnostic spectral features of this series, measured over a range of thicknesses, were used to map similar emulsions in AVIRIS data collected during a relatively calm, nearly cloud-free day on May 17, 2010, using the Tetracorder feature-fitting system. This procedure allowed calculation of oil in emulsions on a per-pixel basis giving 19,000 to 34,000 barrels of oil in the AVIRIS scenes. Based on laboratory measurements, near-infrared photons only penetrate a few millimeters into water-in-oil emulsions; thus, the volume of oil derived using this method is a minimum range. Sheens were too thin to exhibit diagnostic vibrational absorptions, and methods for estimating their volume with AVIRIS data are under development. Because of the great extent of the spill, AVIRIS only covered about 30 percent of the core spill area composed of emulsions and sheens. Extrapolation of AVIRIS-derived emulsion density to the core spill area defined on a MODIS (Terra) image collected the same day indicates a minimum of 66,000 to 120,000 barrels of oil floating on the ocean surface that day. This estimate does not include oil in sheens, oil under the surface, oil washed onto beaches and into wetlands, or oil burned, evaporated, or biodegraded as of May 17. Because of the limited penetration of light into emulsions, and based on field observations that emulsions sometimes exceed 20 millimeters in thickness, we estimate that the oil volume, including oil thicker than can be probed with AVIRIS imagery, is possibly as high as 150,000 barrels in the AVIRIS scenes. Extrapolation of this value to the entire spill gives a possible volume of 500,000 barrels for thick oil remaining on the ocean surface as of May 17.

[8]Clark, R.N., Swayze, G.A., Leifer, Ira, Livo, K.E., Lundeen, Sarah, Eastwood, Michael, Green, R.O., Kokaly, Raymond, Hoefen, Todd, Sarture, Charles, McCubbin, Ian, Roberts, Dar, Steele, Doris, Ryan, Thomas, Dominguez, Roseanne, Pearson, Neil, and the Airborne Visible/Infrared Imaging Spectrometer (AVIRIS) Team, 2010, A method for qualitative mapping of thick oil spills using imaging spectroscopy: U.S. Geological Survey Open-File Report 2010–1101, http://pubs.usgs.gov/of/2010/1101/.

Spatial Association between Pesticide Use and Levels of Obesity and its Co-morbidities in the United States
by Benoit D. Tano

Altru Health System and The University of North Dakota School of Medicine, drbtano@gmail.com

Obesity has become epidemic in America, and the sources appear to be multi-factorial. Recent studies are pointing to endocrine disrupting chemicals as possible culprits. This study was to identify the contribution of pesticides used in farmlands to the obesity epidemic in the United States.

Healthcare Cost and Utilization Project (HCUP) data, the Centers for Disease Control (CDC) obesity maps compiled from data collected through CDC's Behavioral Risk Factor Surveillance System (BRFSS), which each year, reports State health departments data obtained through a series of telephone interviews with U.S. adults, and the U.S. Geological Survey (USGS) pesticide use estimate summarized by county, were used to correlate hypothyroidism, obesity, and selected co-morbidities to pesticide use density. Thyroid-disrupting chemicals such as alachlor, acetochlor, Paraquat, Malathion, and glyphosate and estrogen-disrupting chemicals such as carbaryl, endosulfan, and atrazine, were used in this study. Hospital discharges for hypothyroidism, morbid obesity and selected co-morbidities, and pesticide use density, were analyzed for all U.S. regions. The study focused on years 1997 and 2002, which encompassed all three databases.

Descriptive statistics employed [means (Standard Error of the Mean (SE)), and proportions], as appropriate. A two-tailed p-value < 0.05 denotes statistical significance. When the three databases were analyzed concomitantly, it was found that the thyroid- and estrogen-disrupting chemicals used in this analysis had highest average annual use of active ingredient (pounds per square mile) and cumulative effects in the South, especially in the Mississippi Delta and the Midwest Corn Belt areas. Hypothyroidism discharges for 2002 were also more prevalent in females [3,727 (248), 73%] and in the South [1,952 (150), 38.6%] and the Midwest [1,115 (102), 22%]. The Northeast had [909 (78), 17.99%] and the West had [1,075 (199), 21.3%] for 2002. Statistics for 1997 painted the same picture but were not reported here. Morbid obesity hospital discharges were 74,179 (8,601) total and were also more prevalent in females [62,224 (7,107), 83.9%] and consistent with the high hypothyroidism rate in this group. The regional obesity rates were [20,502 (3,605), 27.6%] for the South, [13,026 (2,965), 17.6%], for the Midwest, [22,844 (5,840), 30.8%] for the Northeast and [17,808 (4,251), 24%] for the West in 2002. The regional rates were not reported by the HCUP in 1997.

These results indicate that endocrine-disrupting chemicals should be scrutinized and regarded with greatest suspicion for contributing to the American obesity-epidemic. Although the areas with highest pesticide use density also seem to have the highest hypothyroidism and obesity rates, this study did not focus on a causality analysis and therefore no cause and effect conclusions are drawn. USGS map application studies should integrate several national databases to arrive at meaningful discoveries and conclusions.

Data Management, 3:45 p.m. – 5:30 p.m.

Demystifying SDE—A Retrospective on the Trials and Tribulations of Implementing ArcSDE as a Master Geospatial Data Library Platform
by Nancy A. Damar[1] and Rose L. Medina[2]

[1]*U.S. Geological Survey Nevada Water Science Center, Henderson Office, 160 N. Stephanie Street, Henderson, NV 89074, (702) 564-4523 nadamar@usgs.gov*
[2]*U.S. Geological Survey Nevada Water Science Center, 2730 N. Deer Run Road, Carson City, NV 89701, (775) 887-7620 rlmedina@usgs.gov*

In 2007, the Nevada Water Science Center began implementation of an Esri ArcSDE/Microsoft SQL Server geospatial data management solution. The content, organizational structure, and even the data format of the Center's master geospatial data library were out of date. At the same time, demand to put data on display and under unprecedented scrutiny through Web mapping applications was increasing for both public information products and internal projects. Additionally, the two large project offices within the Center were maintaining separate, and often different, master libraries. Even without a strong need for a multi-user editing environment, ArcSDE's database and replication capabilities made it the best available tool to address our master library renovations while positioning us to take advantage of future technologies. The following objectives were defined: reorganize data structure, update data content and format from coverage to geodatabase, develop a data management and replication plan, provide a single source of authoritative data, and prepare to develop Web mapping applications. Four years later, though we are not serving any Web maps, the Center has a functioning, replicating SDE-based geospatial data library. This presentation will follow the trials and tribulations of this ongoing project. An overview of the inner workings of the library—hardware setup, data organization, naming conventions, and replication plan—will be discussed, along with some of our missteps and lessons learned on technical issues such as administrator and user account privileges, data loading techniques, and general communication with computer support.

Enterprise Linear Referencing using the High-Resolution National Hydrography Dataset (NHD) and the Hydro Event Management (HEM) Tools: The Oregon Bureau of Land Management (BLM) Experience
by Jay Stevens

Bureau of Land Management, Oregon State, gstevens@blm.gov

Riparian habitat information, fish presence data, and streamflow characteristics are key data elements that inform BLM's planning decisions in the Pacific Northwest. Oregon BLM manages these and many other data elements as linear events in a large (~3.5 million records) database using the National Hydrography Dataset (NHD) as a base route system. Oregon BLM, in cooperation with its national partners, the U.S. Environmental Protection Agency (USEPA) and the U.S. Geological Survey (USGS), has developed a toolset for the creation and management of event data. Under this collaborative approach, BLM provides ongoing development support and USGS provides for tool distribution, primary user support, and training. In this session, we will discuss the use of the HEM tools in the context of Oregon BLM's enterprise event database. Topics will include user experience in working with events using the HEM tools, the database and application architecture, tools and workflows for maintaining events in response to NHD edit activities, and the production of business analytics that support planning activities.

Mapping Out a Course of Action: How the Texas Water Science Center is Using Web-Based Applications to Provide and Manage Geospatial Data, and Lessons Learned along the Way
by Christy-Ann M. Archuleta

U.S. Geological Survey, Texas Water Science Center, 1505 Ferguson Lane, Austin, TX 78754, (512) 927-3516 carchule@usgs.gov

The U.S. Geological Survey (USGS) Texas Water Science Center is increasingly using the Internet for managing and sharing information. In Texas, the USGS collaborates with approximately 120 cooperators and collects a wide variety of hydrologic data,all of which has a geospatial component. In our current (2011) operational model, these data are interpreted and disseminated to our cooperators and the public using different information products. In addition to using online databases and publications to disseminate data and their meaning, we provide many Web-based mapping applications. These Web-based mapping applications are designed for several purposes, including spatial database access and query, data management for storage and retrieval, data rescue for archived materials, and interactive data review. The transition to our current operational model took place over many years, and many lessons were learned along the way. This presentation provides an opportunity to share some of the ways the USGS is using Web-based applications to provide and manage geospatial data in Texas, as well as a chance to recount the history and lessons learned.

Analysis and Modeling, 3:45 p.m. – 5:30 p.m.

Natural-Resources Assessment in Support of Regional Planning and Development in the Anosy Region of Southeastern Madagascar: An Interdisciplinary Geospatial-Based Approach Using Fuzzy Logic
by Mark J. Mihalasky

U.S. Geological Survey, 904 West Riverside Avenue, Spokane, WA 99201, mjm@usgs.gov

U.S. Geological Survey (USGS) scientists from the Mineral Resources, Water, and Ecosystems Programs developed an interdisciplinary, geospatial-based natural-resource assessment methodology, and in 2006, applied the technique to the Anosy Region of southeastern Madagascar. The assessment used expert-knowledge and fuzzy-logic spatial-modeling techniques and involved a geospatial analyst, economic geologists, hydrogeologists, ecologists, an economist, and community and regional development planners. The purpose of the assessment was to enhance knowledge of natural-resource potential in the region and to provide information and decision-making guidance to assist with the creation of a sustainable economic-development model driven by mineral resources. Relationships among geologic/metallogenic, hydrogeologic, ecologic, and socioeconomic data were used (1) to identify priority mineral resource areas and (2) to highlight development growth poles and corridors. Spatial and non-spatial datasets were used to delineate tracts of land permissive for the occurrence of (or known to possess) mineral, water, ecologic, and socioeconomic resources. The tracts were combined, respectively, to create resource-favorability maps. The favorability maps were then integrated with one another, whereby mineral-resource favorability maps were refined by "filtering" them through the favorability maps of the other disciplines. As a result, areas of high mineral-resource favorability were downgraded when proximal to conservation areas or to other regions that might be adversely affected by mineral-resource development. They were upgraded when proximal to roads, water sources, or other infrastructure conducive to development. Similarly, mineral-resource favorability maps were integrated with various tract maps of socioeconomic resources to highlight regions of possible economic and social benefit or detriment that could result from mineral-resource development. The resource-assessment tracts and favorability maps were delineated, combined, and integrated using expert-knowledge and fuzzy-logic spatial-modeling techniques. Tracts were extracted from individual datasets or manually delineated across multiple layers. Resource-favorability maps were generated by combining resource tracts using the fuzzy SUM mathematical operator, which has an overall increasive effect. Resource-favorability maps were integrated with one another using the fuzzy AND mathematical operator, which has a decreasive effect where one or more favorabilities are low.

A 21st Century Conservation Strategy for America's Great Outdoors
by Lisa Duarte,[1] Jocelyn Aycrigg,[2] Anne Davidson,[2] Kevin Gergely,[3] Alexa McKerrow,[4] Adam Radel,[2] Jeff Lonneker,[2] and J. Michael Scott[5]

[1]*National Gap Analysis Program, Stewardship Coordinator, lduarte@uidaho.edu*
[2]*National Gap Analysis Program, University of Idaho, aycrigg@uidaho.edu, adavidson@uidaho.edu, adamr@uidaho.edu, jlonneker@uidaho.edu*
[3]*U.S. Geological Survey, National Gap Analysis Program, 530 Asbury St., Suite 1, Moscow, ID 83843, (208) 885-3565 gergely@usgs.gov*
[4]*U.S. Geological Survey, National Gap Analysis Program, North Carolina Cooperative Fish and Wildlife Unit, North Carolina State University Department of Zoology, Campus Box 7617, Raleigh, NC 27695, (919) 513-2853 alexa_mckerrow@ncsu.gov*
[5]*U.S. Geological Survey, Idaho Cooperative Fish and Wildlife Research - University of Idaho, mscott@uidaho.edu*

While the conservation community has embraced a landscape scale approach for decision making through initiatives like America's Great Outdoors, lack of relevant and consistent data at a national scale has been an impediment for a national conservation strategy. That has changed with the recent release of the U.S. Geological Survey (USGS) Gap Analysis Program's (GAP) national conservation databases, including land cover, protected areas, and vertebrate species ranges and distributions. GAP's National Land Cover Data contains vegetation communities at three nested hierarchical levels with detailed descriptions of each land cover type. The finest level of the hierarchy contains 590 different land cover and land use types. Information regarding the ownership, management, and biodiversity conservation status of Federal, State, city, county, and private conservation lands is in the Protected Areas Database of the United States (PAD-US). The PAD-US is currently the U.S. authoritative data source for the World Database on Protected Areas (WDPA) and is included in the current release of the North American Terrestrial Protected Areas Database. Seasonal ranges and predicted distribution models for approximately 2,000 species across the United States provide valuable information regarding where species occur. Endangered, threatened, and rare as well as Species of Greatest Conservation Need (SGCN) are included. Each database is an effective conservation tool at national (that is, Landscape Conservation Cooperatives), regional, and State scales developed with States, universities and non-governmental organizations. However, combining them provides invaluable information for a conservation strategy that identifies high biodiversity areas, ecologically valuable lands, species underrepresented within our current protected areas network, and stewardship responsibilities. These national data and analyses not only support America's Great Outdoors Initiative but enhance our ability to conserve our wild legacy.

GIS to the Rescue: Saving the Rio Grande Silvery Minnow
by Daniel K. Pearson and J. Bruce Moring

U.S. Geological Survey, Texas Water Science Center, 1505 Ferguson Lane, Austin, TX 78754, (512) 927-3561 dpearson@usgs.gov, (512) 927-3585 jbmoring@usgs.gov

In 2010, the U.S. Geological Survey (USGS), in cooperation with U.S. Fish and Wildlife Service (FWS), began to assess the relation of seasonal flow conditions to available habitat, distribution, and recruitment of reintroduced Rio Grande silvery minnow (*Hybognathus amarus*) in the Big Bend reach of the Rio Grande River in Texas. The FWS, with the assistance of many partners including The National Park Service and Texas Parks and Wildlife Department, released approximately 445,000 silvery minnows at four sites in the Big Bend reach of the Rio Grande River on December 16 and 17, 2008. This was the first of three releases completed to date (January 2011) and marked an important step in fulfilling the goals identified in the Rio Grande Silvery Minnow Recovery Plan (a 10[(j)] nonessential experimental population designation, published on December 8, 2008). Work completed as part of Phase I of the USGS-FWS study (fiscal year 2010) included detailed field mapping of habitat supportive of the silvery minnow using high accuracy GPS in concert with GIS. Acquired data were stored in a geodatabase and presented using an on-line mapping application. Mapped units (mesohabitats) will include comprehensive fish assemblage, physical habitat, and explanatory spatial variables. Results from this study will help to refine the process of release site selection, assist in the development of a more focused species monitoring assessment strategy, and provide detailed physical habitat information for the species over a range of flow conditions.

Wednesday, May 11, 2011, Poster Session, 6:00 p.m. – 8:00 p.m.

Scanning and Georeferencing Historical USGS Quadrangles
by Greg Allord

U.S. Geological Survey, 465 Science Drive, Suite A, Madison, WI 53711, gjallord@usgs.gov

The U.S. Geological Survey (USGS) Historical Quadrangle Scanning Project (HQSP) is scanning all scales and all editions of approximately 250,000 topographic maps published by the USGS since the inception of the topographic mapping program in 1884. This scanning will provide a comprehensive digital repository of USGS topographic maps, available to the public at no cost. When physical and cultural features change over time, maps are updated, revised and new editions printed. Although they are out of date, these historic maps are often useful to scientists, historians, environmentalists, genealogists, and others researching a particular geographic location or area. A series of maps of the same area published over a period of time can show how some areas looked as early as 1884, before current development, and provide a detailed view of changes over time. Historical maps are stored in a limited number of collections and are not readily available to the public. The USGS has begun to convert these historical printed topographic quadrangles to an electronic format. This project serves the dual purpose of creating a master catalog and digital archive copies of the irreplaceable collection of topographic maps in the USGS Reston Map Library as well as making the maps available for viewing and downloading. This presentation will describe how the HQSP is accurately cataloging and creating metadata to accompany high-resolution, georeferenced digital files representing the lithographic maps. Each map image is scanned as is and captures the current content and condition of each map sheet. The project provides ready access to maps that are either no longer available for distribution in print or are being replaced by the new generation of US Topo maps. Georeferencing of the map files that is, tying them to a known earth coordinate system, enables them to be imported into Geographic Information Systems so that they can be overlain with other geospatial (map) data from other sources, such as from *The National Map*. The potential for research that analyzes change over time is becoming increasingly recognized by the geospatial community, and this project will provide published lithographic USGS maps in georeferenced digital formats. With georeferencing, the historical maps can be combined with current data from *The National Map*. The product will be delivered in a compressed GeoPDF format and as GeoTIFF images with embedded metadata.

Mapping Out a Course of Action: How the Texas Water Science Center is Using Web-Based Applications to Provide and Manage Geospatial Data, and Lessons Learned along the Way
by Christy-Ann M. Archuleta

U.S. Geological Survey, Texas Water Science Center, 1505 Ferguson Lane, Austin, TX 78754,
(512) 927-3516 carchule@usgs.gov

The U.S. Geological Survey (USGS) Texas Water Science Center is increasingly using the Internet for managing and sharing information. In Texas, the USGS collaborates with approximately 120 cooperators and collects a wide variety of hydrologic data,all of which has a geospatial component. In our current (2011) operational model, these data are interpreted and disseminated to our cooperators and the public using different information products. In addition to using online databases and publications to disseminate data and their meaning, we provide many Web-based mapping applications. These Web-based mapping applications are designed for several purposes, including spatial database access and query, data management for storage and retrieval, data rescue for archived materials, and interactive data review. The transition to our current operational model took place over many years, and many lessons were learned along the way. This presentation provides an opportunity to share some of the ways the USGS is using Web-based applications to provide and manage geospatial data in Texas, as well as a chance to recount the history and lessons learned.

NHD in the Real World, Features of the Blue River Subbasin: Using Data from *The National Map*
by Ariel Bates[1] and Jeff Simley[2]

[1]*U.S. Geological Survey, TNM Data Operations Branch, Hydrography Section, Box 25046, Denver Federal Center, MS 510, Denver, CO 80225, (303) 202-4535 atbates@usgs.gov*
[2]*U.S. Geological Survey, National Geospatial Program, Core Science Systems, Box 25046, Denver Federal Center, MS 510, Denver, CO 80225, (303) 202-4131 jdsimley@usgs.gov*

The National Hydrography Dataset (NHD) is a comprehensive set of digital spatial data that contains information about surface water features such as lakes, ponds, streams, rivers, springs and wells. Within NHD, surface water features are combined to form "reaches," which provide the framework for linking water-related data to NHD surface water drainage network. These linkages enable the analysis and display of these water-related data in upstream and downstream order.

This map presents the NHD features of the Blue River subbasin. The intent of the map is to show the hydrography features of the Blue River subbasin and how NHD data appear in the real world. By reviewing features it is possible to better understand the real world applications and limitations of NHD data.

NHDPlus Streamflow: Integrating Data from *The National Map*
by Ariel Bates

U.S. Geological Survey, TNM Data Operations Branch, Hydrography Section , Box 25046, Denver Federal Center, MS 510, Denver, CO 80225, (303) 202-4535 atbates@usgs.gov

Various data from *The National Map* can be integrated and synthesized into scientific information that can be used for modeling and analysis. The NHDPlus is one example of how this data integration takes place. Using the National Hydrography Dataset (NHD), the Watershed Boundary Dataset (WBD), the National Elevation Dataset (NED), and streamgages, the NDHPlus estimates values such as streamflow and many other hydrologic statistics. These statistics provide information that can be used for scientific analysis. A few examples of how these data can be utilized include mapping nitrogen depositions, examining impaired waters, assessing watersheds, and protecting drinking water. This poster highlights how *The National Map* data are integrated to create the NHDPlus, and, more specifically, streamflow statistics to contribute to national, State and local agencies' scientific needs.

Wyoming's Stewardship Efforts to Fix GNIS Issues within the NHD
by Paul Caffrey,[1] Pat Madsen,[2] and Joel Skalet[3]

[1]*GIS Research Scientist/Education Coordinator, University of Wyoming, Wyoming Geographic Information Science Center (WyGISC), Dept. 4008, 1000 E. University Avenue, Laramie, WY 82071, (307) 766-2770 (office) (307) 766-2744 (fax) caffrey@uwyo.edu*
[2]*GIS Research Scientist, Wyoming Geographic Information Science Center (WyGISC), Dept. 4008, 1000 E. University Avenue, Laramie, WY 82071, (307) 766-2770 patsmad@uwyo.edu*
[3]*U.S. Geological Survey, NHD Partner Support Cartographer, NGTOC, 505 Science Drive, Madison, WI 53711-1061, (608) 238-9333 ext. 117 jjskalet@usgs.gov*

As rich and great as the National Hydrography Dataset (NHD) is, it is not perfect. There will always be a need to improve or enhance the data for the benefit of its users. The Geographic Names Information System (GNIS) data were used to populate the stream names in the NHD as well as many of the hydrologic units included in the Watershed Boundary Dataset (WBD) for the Nation. In general, the process of assigning names and GNIS ID numbers to NHD features was very complete. However, discrepancies persist in every region of the country, in part, due to inherent GNIS line vector coordinate errors as well as interpretative mistakes made during GNIS conflation process when creating the dataset. Under funding provided from the 2009 U.S. Geological Survey (USGS) Partnerships Opportunity Grants, the Wyoming Geographic Information Science Center (WyGISC) has been working with the USGS in an effort to improve and update the National Hydrography Dataset (NHD) and Watershed Boundary Dataset (WBD) for the State of Wyoming. Primary goals for this project are to improve integration of the Geographic Names Information System (GNIS) data with the NHD by updating missing, misplaced, and misspelled GNIS names and the incorrect GNIS feature IDs for the NHD's stream and waterbody features. This presentation will provide others who are now working to improve GNIS named features in the NHD our experience in undertaking the task to correct GNIS errors and to improve GNIS name attribution for major water features within the NHD for the State of Wyoming.

Updating the LASED Map: Creating a Modern Web Mapping Interface
by Brendan Dwyer

U.S. Geological Survey St. Petersburg Coastal and Marine Science Center, 600 4th Street South, St. Petersburg, FL 33701, (727) 803-8747 ext. 3123 bdwyer@usgs.gov

The Louisiana Sedimentary and Environmental Database (LASED) is a cooperative effort to organize and disseminate geologic information for the Louisiana coastal zone. In 2005, the project created an Esri Internet Mapping Service (IMS) Web site to serve these data to the public over the internet. Although routine patches and software upgrades were applied, by 2009, the hardware and IMS software used to support the site were slow and outdated. The St. Petersburg Coastal and Marine Science Center Web mapping services team recently planned and executed an upgrade from IMS to an Esri ArcGIS Server system. While IMS offers spatial data viewing and querying functionality, the older technology was slow to respond to user actions and required an entire map redraw any time the user panned or zoomed in or out. Esri ArcGIS Server is a modern set of software and technologies for storing and serving spatial data, and offers additional functionality such as editing over the network and geoprocessing. The hardware was upgraded from a Sun Fire server running Solaris to an IBM xSeries host running Red Hat Enterprise Linux, version 5. ArcGIS Server was used to create several Web services to deliver the LASED spatial layers. We chose to use the ArcGIS Viewer for Flex template as a Web-mapping interface and modified it to suit our needs. Flex is a software development platform from Adobe. The ArcGIS Flex Viewer is a customizable Flex Web-mapping application that developers can configure and extend. We extended the functionality of selected features of the template. For example, we modified the object that displays records returned from queries. The modifications dictate the displayed order of columns from the table, omit a column if it does not contain a value, and create a button that acts as a hyperlink if the column contains a URL. We used the Esri GIS services basemaps, which freed us from making and supporting our own basemap layers. Using the Esri basemaps made the interface faster and easier to maintain. The new map application is a "slippy map," like GoogleMaps, that allows dynamic panning and zooming without an entire page redraw. The interface includes an identify tool and a query function that allow users to query the attribute tables of the layers. The new site is easier to use and is more intuitive. The result is a modern Web mapping application that is faster and easier to maintain than was IMS.

Texas-Sized Geodatabases of Watershed Characteristics: Applying GIS in Support of Water-Quality Remediation
by Sophia Gonzales, Christy-Ann Archuleta, and David R. Maltby

U.S. Geological Survey Texas Water Science Center, 1505 Ferguson Lane, Austin, TX 78754,
(512) 927-3507 slhurtad@usgs
(512) 927-3516 carchule@usgs.gov
(512) 927-3588 drmaltby@usgs.gov

The U.S. Geological Survey, in cooperation with the Texas Commission on Environmental Quality, is automating the delineation of watershed boundaries and compiling information pertaining to watershed characteristics for more than 3,000 surface-water monitoring stations throughout Texas. The Texas Commission on Environmental Quality collects and analyzes water-quality data at these stations to ensure compliance with applicable State and Federal water quality standards. When the water quality at any given station fails to meet applicable standards, a plan for remediation is developed. The remediation plans require detailed watershed boundary delineation and characterization of spatial data including precipitation statistics, elevation, slope, land use or land cover statistics, and total drainage area. Visual Basic scripts utilizing ArcObjects1 are used to automate the watershed delineation and characterization calculations processes, and a comprehensive geodatabase of each watershed's characteristics is developed. The geodatabases are designed to facilitate remediation and resource management decision-making through the visualization of geographic information and the integration and analysis of spatial data.

Landsat 5 Data for Alaska
by Tom Heinrichs, Dayne Broderson, and Jay Cable

University of Alaska Fairbanks - Geographic Information Network of Alaska – Director tom.@alaska.edu,
dayne@gina.alaska.edu, jay@gina.alaska.edu

Landsat 5 is an Earth-imaging satellite with many applications, including resource management, crisis response, and hazard monitoring. There is not a functioning onboard recorder on Landsat 5, so satellite data must be transmitted directly to Earth as it images the ground beneath the satellite. From 1996 to 2005, very few Landsat 5 data were collected over Alaska. Following the failure of the Landsat 7 scan line corrector in 2003, the need for Landsat 5 data for Alaska became urgent. In 2005, the U.S. Geological Survey (USGS) National Center for Earth Resources Observation Systems (EROS), NOAA/NESDIS Fairbanks Command and Data Acquisition Station (FCDAS), and University of Alaska Fairbanks' Geographic Information Network of Alaska (GINA) teamed up to provide Landsat 5 data for Alaska. Between 2005 and 2011 more than 21,000 Landsat 5 scenes were captured at FCDAS and transmitted over GINA's network to EROS for processing and archiving. These data have proved vital to scientists, resource managers, and emergency responders and would not have been available without this multi-agency collaboration.

GIS-based Integration of New Geologic Mapping, Satellite-derived Quartz Mapping, and a Statewide Digital Geologic Map Yields Insights into the Roberts Mountains Allochthon Applicable to Assessments for Concealed Carlin-type Gold Deposits
by Chris Holm-Denoma,[1] Albert H. Hofstra,[2] Barnaby W. Rockwell,[3] and A. Elizabeth Jones Crafford[4]

[1]U.S. Geological Survey, Central Mineral and Environmental Resources Center, Box 25046, Denver Federal Center, MS 973, Denver, CO 80225, (303) 236-5454 cholm-denoma@usgs.gov
[2]U.S. Geological Survey, Central Mineral and Environmental Resources Center, Box 25046, Denver Federal Center, MS 973, Denver, CO 80225, (303) 236-5530 ahofstra@usgs.gov
[3]U.S. Geological Survey, Central Mineral and Environmental Resources Center, Box 25046, Denver Federal Center, MS 973, Denver, CO 80225, (303) 236-1851 barnabyr@usgs.gov
[4]GeoLogic, 9501 Nettleton Dr., Anchorage, AK, 99507

The Great Basin physiographic province is the Nation's premier gold mining region. Most of the gold currently produced is from Eocene Carlin-type gold deposits (CTDs) hosted in permeable reactive lower Paleozoic miogeoclinal carbonate rocks exposed in erosional windows through less permeable coeval eugeoclinal siliciclastic rocks of the Roberts Mountains allochthon (RMA). The RMA was assembled in a regionally extensive fold and thrust belt during the early Mississippian Antler and Permian Humboldt orogenies (Holm-Denoma and others, in review). Most CTDs exposed in windows have been found, such that current exploration involves drilling to depths of 1-2 kilometers through adjacent areas of the RMA. An improved understanding of the thickness and internal structure of the RMA would benefit exploration and assessments for concealed CTDs. The RMA is composed of complexly deformed relatively monotonous successions of interbedded shale, chert, and quartzite with minor greenstone and limestone, which has made it difficult to document stratigraphic and structural relationships. In the Independence Mountains of northern Nevada, we have identified a coherent marker, the thick-bedded quartzite of the Valmy Formation (Ovq), which is in thrust fault contact with quartz-poor successions in the RMA. The thrust fault below the Ovq is an important datum because the overall thicknesses of RMA between it and underlying CTD host rocks is typically <600 meters, and in some locations, <100 meters. The ASTER (Advanced Spaceborne Thermal Emission and Reflection Radiometer) sensor aboard the Earth Observing System Terra satellite acquires multispectral image data, including five bands from the thermal region (8-14 μm) of the electromagnetic spectrum. The thermal infrared ASTER data are capable of detecting both hydrothermal hydrous silica and nonhydrous varieties of quartz. Quartz mineral maps of northern Nevada have been used to identify thick sequences of quartz-rich sedimentary and metamorphic rocks and hydrothermal silicification in carbonate rocks near CTDs (Rockwell and Hofstra, 2008). In the Independence Mountains, Ovq stands out on the regional ASTER-based quartz map. To establish the regional extent of the Ovq thrust sheet, remote sensing was used to locate additional exposures of Ovq in areas previously mapped as RMA in the digital geologic coverage of Crafford (2007). A number of other areas along a 200-kilometer trend were identified and investigated in the field that revealed similar relationships to those in the Independence Mountains. These are areas where the RMA is sufficiently thin for potential CTDs to be in reach of drilling.

Water Resources Discipline (WRD) National Spatial Data Infrastructure (NSDI) Node
by Michael C. Ierardi

U.S. Geological Survey, 12201 Sunrise Valley Drive, MS 445, Reston, VA 20192, (703) 648-5649 mierardi@usgs.gov

The Water Resources NSDI Geospatial Node is an online centralized repository of digital data conforming to the Federal Geographic Data Committee (FGDC) XML metadata standards, methods of data acquisition, and processing protocols that are required in the FDGC's Content Standard for Digital Geospatial Metadata (CSDGM). The valid metadata schema on the WRD NSDI Geospatial Node provides an organized structure of categories and essential elements describing the identification of data quality, publishing organization, references, entity and attribute information, and distribution methods for each spatial dataset. The author has the option of appending additional metadata citations such as an abstract, purpose, supplemental information, graphics, techniques applied in data base development, findings of the investigations, project summaries, themes, keywords, or other citations the author determines are relevant or appropriate to the dataset.

The WRD node contains over 748 metadata files referencing over 3,325 downloadable geospatial datasets in compressed formats. After a dataset is downloaded to a user's computer, these accessible geospatial files can be readily displayed by Geographic Information Systems (GIS). The WRD NSDI Node utilizes online Web Accessible Folder (WAF) technology to archive and list GIS datasets. The published catalog of data holdings on the WRD NSDI Geospatial Node is centralized in a repository on the USGS Water Discipline server, where its metadata are harvested by Geospatial One-Stop (http://gos2.geodata.gov/wps/

portal/gos) through their textual and spatial schema, which also creates a geographical footprint or geographic outline for each dataset, allowing the files to be cited, searched, retrieved, and downloaded.

Geoportals allow geospatial information to be accessed through indexes or catalogs, and further help other agencies to conserve their GIS resources by reducing overhead costs from duplicating existing data efforts. With the increase of data availability, the benefits of utilizing past and present data will help foster partnerships within nations, States, counties, and agencies. To prepare the WRD NSDI Node for the International Organization for Standards metadata standard (ISO 19115), a new set of core metadata elements needs to be populated for each metadata set. These mandatory elements of the ISO core metadata standard require additional information that cannot be directly captured from the data. Therefore, the core elements must be mapped to existing CSDGM metadata elements and appended into the metadata using some form of a data integrated metadata tool. Spatial datasets are available on the WRD NSID Geospatial Node at http://water.usgs.gov/lookup/getgislist.

Integration of the Watershed Boundary Dataset and The National Hydrography Dataset
by Kathy Isham[1] and Stephen Daw[2]

[1]U.S. Geological Survey, Box 25046, Denver Federal Center, MS 510, Denver, CO 80225, (303) 202-4419 krisham@usgs.gov
[2]U.S. Geological Survey, Box 25046, Denver Federal Center, MS 510, Denver, CO 80225, (303) 202-4418 sgdaw@usgs.gov

The Integration of the Watershed Boundary Dataset (WBD) with the National Hydrography Dataset (NHD) was an important milestone for the NHD. The WBD was originally maintained by the U.S. Department of Agriculture, but the U.S. Geological Survey (USGS) took over the WBD upon the release of NHD Model 2.0 in 2010. The watershed boundaries included in the dataset range from the largest units (Regions),with a 2 digit hydrologic unit codes, to the smallest units, with a 16-digit hydrologic unit code. Watershed Boundaries delineate the drainage extent of surface-water systems. This poster illustrates how the WBD can be combined with the NHD to understand how surface water exits each watershed boundary at a single point. As streams from watersheds converge, the volume of water generally increases until the waters finally reach the end of a system such as at a coastline. Another important theme is that water does not follow manmade boundaries such as country borders. At this time, the USGS is working on incorporating watershed boundaries that cross the international borders of Canada and Mexico.

Quick Reference Guide for NHD Features
by Kathy Isham[1] and Charles Bowker[2]

[1]U.S. Geological Survey, Box 25046, Denver Federal Center, MS 510, Denver, CO 80225, (303) 202-4419 krisham@usgs.gov
[2]U.S. Geological Survey, Box 25046, Denver Federal Center, MS 510, Denver, CO 80225, (303) 202-4504 crbowker@usgs.gov

The National Hydrography Dataset (NHD) is the surface-water component of *The National Map*. The NHD is a digital vector dataset used by geographic information systems. The data contained in the NHD are used in general mapping and analysis of surface-water systems. The Quick Reference Guide for NHD Features provides a description of the different features and feature classes within the dataset, defines how these features relate to real-world hydrography, and also details how they may be used in geographic analysis. This helps GIS users better understand how surface-water features on the landscape equate to the lines, points, and polygons in the NHD.

Canadian-U.S. Hydrographic Data Harmonization and Integration
by Michael T. Laitta

International Joint Commission of Canada and the U.S., Physical Science Advisor, Physical Scientist, GIS Coordinator, 2410 Pennsylvania Ave. NW, Washington, DC 20037, (202) 736-9022 (202) 341-1487, laittam@washington.ijc.org

The International Joint Commission (IJC), in coordination with Environment Canada, Natural Resources Canada, U.S. Geological Survey, Agriculture and Agri-Foods Canada has made substantial progress with the harmonization of the shared fundamental hydrographic datasets along the Canadian-U.S. interface. Phases I and II of this effort, the alignment and editing of sub-drainage areas within the major Transboundary Basins and the first pass connection of the fundamental hydrographic layers (the U.S. National Hydrography Dataset [NHD] and the Canadian National Hydro Dataset [NHN]), are close to completion. In 2011, this effort will focus on the delineation and refinement of smaller drainage units within these now-harmonized sub-drainage areas—the extension of the U.S. Watershed Boundary Dataset (WBD) into Canadian territory. Coincidental to this next step, the

IJC is encouraging the development/expansion of binational water quality and quantity applications such as StreamStats and the U.S. SPARROW model. This presentation will touch upon the basic technical methods employed to facilitate the negotiation of binational delineations, impacts to the Federal stewarding agencies, and potential opportunities for sustainable, regionally based yet binational hydrologic applications for the shared U.S.-Canadian geographic interface.

The National Hydrography Dataset (NHD)
by Tony Litschewski[1] and Ariel Bates[2]

[1]*U.S. Geological Survey, Box 25046, Denver Federal Center, MS 510, Denver, CO 80225,*
(303) 202-4292 aalitschewski@usgs.gov
[2]*U.S. Geological Survey, TNM Data Operations Branch, Hydrography Section, Box 25046, Denver Federal Center MS 510, Denver, CO 80225, (303) 202-4535 atbates@usgs.gov*

The National Hydrography Dataset (NHD) is the surface-water component of *The National Map*. The NHD is an important digital spatial data layer to the GIS community at local, State, Federal, and international levels. The NHD is used by private and government entities for data modeling, emergency response, and surface water analysis. The NHD is maintained by the U.S. Geological Survey (USGS) and updated by the USGS as well as many national partners. NHD data are open and freely accessible to all. In the future the NHD will continue to play a vital role not only in mapping but also in scientific analysis and modeling of hydrologic features.

Using the Landsat Archive in Court
by Melinda McGann

U.S. Forest Service, Rocky Mountain Region, Remote Sensing Specialist, mlmcgann@fs.fed.us

The U.S. Forest Service used archived Landsat scenes to prove a timber trespass in court. Since then, the U.S. Department of Justice asked for the same support in a multi-jurisdiction water-rights case. This presentation and poster show how we searched the archive, downloaded and processed the scenes, and displayed the images in the courtroom.

Powell and the Watershed Boundary Dataset
by Jeffrey D. Simley

U.S. Geological Survey, National Geospatial Program, Core Science Systems, Box 25046, Denver Federal Center, MS 510, Denver, CO 80225, (303) 202-4131 jdsimley@usgs.gov

If Congress had listened to explorer and scientist John Wesley Powell 125 years ago, the American West today might be an entirely different place. In 1878, Powell published his Report on the Lands of the Arid Region, which laid out a concrete strategy for settling the West without fighting over scarce water. Powell later produced a map of the arid region of the United States showing drainage districts, 1890-91. This map would lay the groundwork for Powell's suggestion to organize the West's political boundaries on the basis of his drainage districts. With the completion of the Watershed Boundary Dataset for the United States in 2008, the opportunity now exists to compare Powell's 1890 map with contemporary mapping using modern analysis techniques. This comparison provides insight into Powell's techniques and reasoning, and tells us much about what might have happened if Powell's proposal for organization of the Western lands had been adopted.

St. Petersburg Coastal and Marine Science Center's Core Archive Portal
by Matthew Streubert, Chris Reich, and Brendan Dwyer

U.S. Geological Survey, St. Petersburg Coastal and Marine Science Center, 600 4th Street South, St. Petersburg, FL 33701, (727) 803-8747 ext. 3044, mstreubert@usgs.gov, creich@usgs.gov, bdwyer@usgs.gov

The St. Petersburg Coastal and Marine Science Center (SPCMSC) houses a unique collection of coral and rock cores from around the world. Archived cores consist of 3- to 4-inch-diameter coral cores, 1- to 2-inch diameter rock cores, and a few unlabeled loose coral and rock samples. Many of the cores in the collection are from south Florida, but there are samples from

Enewetak Atoll, the Philippines, the U.S. Virgin Islands, Belize, the Caribbean, and the Gulf of Mexico to name a few. The primary product of this work is the Web portal that allows the user to search for core material via a world map. The user follows the Web link to access the core archive database. In order to prepare the core material data and present it in Web format, the database is first encoded in the scripting language Hypertext Preprocessor (PHP). Once the database is coded in PHP, it is linked to a database management system, the open-source MySQL software platform. The archive curator has the capability to access the database via a Web interface, which allows for the modification of the archive records, while the external user will only be able to view the display and download core material information. The core archive database is presented in two different formats: (1) a table-driven PHP page and (2) a searchable map through ArcGIS Server 10. The user can search the table-driven PHP page by entering query information into the search field. Eventually, the query field will have advanced features that will enable a user to search fields using specific terms. The ArcGIS Server platform will allow for the core archive data to be served up using Flex Viewer, which is a Flash-driven mapping interface. The interface gives the user the ability to identify, query, and download archive records from a map interface. The map interface is not limited to the functions above and offers the user more than just a map view of the location of the core archive records. The user will be able to zoom in on an area of interest and click on a bullet that marks the core location. The site and core information for the selected core will also appear on the screen. The user has control on whether to download the information for that particular core site or to zoom into a particular area and download the entire archive list for that specified region. The user also can even specify the format of the download file.

Utilizing Image Metadata Headers to Store Geospatial Information
by Matthew Streubert

U.S. Geological Survey, St. Petersburg Coastal and Marine Science Center, 600 4th Street South, St. Petersburg, FL 33701, (727) 803-8747 ext. 3044 mstreubert@usgs.gov

Digital images have the ability to convey much more information than simply the pixels displayed on the screen. At the St. Petersburg Coastal and Marine Science Center, we use that capability to make a smarter database of images and to use those photos in conjunction with GIS software. Along with the pixel information, each image contains headers that can be edited on-the-fly or after an image is acquired. Header types such as EXIF, IPTC, XMP, and others, can be post-populated to store information such as artist, copyright, location, comments, State, and more. There are over 300 fields available to populate in the various headers of JPEGs and TIFFs. With so many available fields, GPS-enabled cameras can store spatial information including position, elevation, and direction. Alternatively, a user can populate any field available with software such as EXIFutils, an EXIF-metadata populator, without needing a GPS-enabled camera. Image headers can be read using PHP, Javascript, Perl, C#, and possibly other coding platforms. Software applications such as Flickr, Firefox, Google Maps, Google Earth, and ArcPhoto can then read a number of fields and provide a user with information about the image and its spatial reference. This information, as long as the image is not altered, remains with the picture for its life; in fact, it becomes the image metadata. Tools to generate image metadata are readily available, cost efficient, and in some cases, even free. It is our local mission to ensure that information is stored for all our photo collections within our photodatabase, providing end users with photo metadata. Our photodatabase and accompanying Web site were created using a MySQL database, PHP, and Apache. It utilizes a PHP script (Exifer) to automatically extract the image headers and populate the MySQL database. This technique allows us to extract all image metadata from the image and to filter it into corresponding fields, and then use it for querying. Currently, the database is home to over 80,000 images with over 10,000 on deck.

Refining Subsurface Geologic Framework of the Amargosa Desert, Nevada-California, Using 3-D GIS and Geologic Framework Modeling
by Emily M. Taylor and Donald S. Sweetkind

U.S. Geological Survey, Box 25046, Denver Federal Center, MS 980, Denver, CO 80225,
(303) 236-8253 emtaylor@usgs.gov
(303) 236-1828 dsweetkind@usgs.gov

The U.S. Geological Survey (USGS) is currently involved in an interdisciplinary project constructing a detailed numerical 3-D framework and groundwater flow model of the Amargosa Desert, NV-CA, embedded within the existing USGS Death Valley regional numerical groundwater flow model. The goal of the Southern Amargosa Embedded Groundwater Flow Model (SAMM) is to address Department of the Interior agencies' concerns regarding groundwater use in the Amargosa related to population growth and proposed solar energy production. We are using 3-D GIS and geologic framework modeling techniques to refine the subsurface geologic framework of the Amargosa Desert basin; results will form the geologic inputs to the numerical groundwater flow model. We have developed a 3-D GIS database of lithologic, stratigraphy, and geophysical data from 466 boreholes from various sources including water wells, mining exploration, and USGS and Nye County, NV, monitoring wells. Complex driller's lithologic descriptions were assigned to 30 lithologic classes based on interpretation of lithologic and sedimentologic characteristics. Stratigraphic tops were interpreted from lithologic sequences and correlations to Cenozoic geologic units mapped at the surface. We use the 3-D GIS database as the basis for numerical extrapolation of geologic data within a 3-D geologic framework model using RockWorks 15 modeling software. The SAMM geologic model is 67 by 81 kilometers with a grid spacing of 500 meters in the x and y directions with 10-meter z spacing. The subsurface geologic model reflects the interpreted 3-D distribution of lithotypes, associated stratigraphic units, and the location of basin-bounding and intrabasin structures. Faults within the basin have sufficiently large normal and strike-slip offset to segment the basin fill into discrete lithologic and stratigraphic packages. We constructed the multiple fault-bounded 3-D lithologic models by extracting information from our 3-D GIS data that occurred within fault-bounded domains. The final 3-D geologic model is created by numerically adding together six fault-bounded submodels. Our intention was to preserve the geologic structures that could potentially affect groundwater flow. From the lithologic model, nine stratigraphic boundaries were created. Both lithologic and stratigraphic characteristics will be considered in the hydrologic model.

Stream Ecological Condition Modeling at the Reach and the Hydrologic Unit Code (HUC) Scale: A Look at Model Performance and Mapping Uncertainty
by Marc Weber and John Van Sickle

U.S. Environmental Protection Agency, National Health and Ecological Effects Research Laboratory, Western Ecology Division, Corvallis, OR, (541) 754-4314 weber.marc@epa.gov, vansickle.john@epa.gov

The National Hydrography and updated Watershed Boundary Datasets provide a ready-made framework for hydrographic modeling. Determining particular stream reaches or watersheds in poor ecological condition across large regions is an essential goal for monitoring and management. To address this need, we built predictive models of stream ecological condition (water quality, biological condition, and physical habitat) across the Pacific Northwest (PNW; Oregon, Washington, and Idaho) using the enhanced National Hydrography Dataset (NHDPlus) as the spatial framework. Using data from the U.S. Environmental Protection Agency (USEPA) EMAP-West survey, from the U.S. Geological Survey (USGS) National Water-Quality Assessment (NAWQA) Program, and from other regional monitoring efforts, we constructed a number of predictive models for several stream condition endpoints using 136 landscape metrics as predictor variables. We mapped these model results to every NHDPlus stream reach in the PNW, aggregated predictions across all reaches within each 6th-level USGS Hydrologic Unit Code (HUC), and mapped the HUC-averaged scores. We examine the substantial uncertainties of our predictions and suggest how they could be reduced, as well as address issues of aggregating results to coarser spatial scales (from individual stream reaches to watersheds or HUCs). We also look at differences in our models based on aggregation of stream reach results for only headwater streams in a given hydrologic unit versus results for streams flowing through multiple hydrologic units.

The Role of Hydrologic Information in the Conservation, Management, and Restoration of National Wildlife Refuges in the Southeastern United States
by Loren Wehmeyer[1] and Gary Buell[2]

[1]*U.S. Geological Survey, Texas Water Science Center, 1505 Ferguson Lane, Austin, TX 78754, (512) 927-3568 llwehmey@usgs.gov*
[2]*U.S. Geological Survey, Georgia Water Science Center, 3039 Amwiler Road Suite 130, Atlanta, GA 30360, (770) 903-9160 grbuell@usgs.gov*

Ensuring an adequate water supply is a growing issue for many U.S. Fish and Wildlife Service (USFWS) National Wildlife Refuges in the Southeastern United States. Understanding the past and present hydrologic regime is a prerequisite for establishing quantifiable metrics of water demand in relation to water availability for ecological needs. Accordingly, the U.S. Geological Survey (USGS), in cooperation with the USFWS, is investigating the role of hydrologic information in the conservation, management, and restoration of National Wildlife Refuge biological resources and critical habitats. Providing compiled hydrologic information to local, State, and Federal governments during the legislative process can help ensure that the ecological needs of National Wildlife Refuges are considered prior to the determination of dam re-licensing, minimum in-stream flow requirements, groundwater and lake level requirements, and other items concerning water management. Under Federal and State laws, the USFWS has the legal authority to secure water rights to ensure refuges have adequate water supplies of sufficient quality. The ability to negotiate for future water resources in an era of increased competition among users depends on legally defensible, objective, and quantifiable data. The work presented here provides watershed characterization data (raster and feature class geo-databases of such diverse information as aerial photographs, land use, geology, and locations of roads, levees, and streamflow-gaging stations). Also included is location-specific documentation of past and present water quantity, including the timing and distribution of flow in streams providing water for selected high priority refuges in the Southeast. Future water supplies for these refuges are uncertain because of known or potential stressors such as droughts, population growth, hydrologic manipulation, ongoing litigation, or recent or proposed changes in water laws.

Thursday, May 12, 2011, *The National Map* Users Conference Plenary Sessions, 8:30 a.m. – 10:30 a.m.

Host: Kevin Gallagher

Speakers:

Mark DeMulder

Current Status of *The National Map* and a Vision for the Future

The National Map, the cornerstone of the U.S. Geological Survey's National Geospatial Program, is a collaborative effort between the U.S. Geological Survey and other Federal, State, and local partners to improve and deliver geospatial information for the Nation. This plenary talk will describe the current status of *The National Map* and a vision for the future. Forging and engaging in new relationships, responding to users' needs, and leveraging partners and public knowledge are all part of our future direction. The nexus between social media and geospatial technology will bring exciting opportunities for moving toward this goal.

Deanna Archuleta

Advancing our Geospatial Foundation for Protecting America's Great Outdoors and Powering Our Future

Easily accessible, standardized geospatial data are essential in providing a common geographic reference to all citizens, in spurring commercial development of geospatial tools, and in carrying out the complex missions entrusted to the Department of the Interior. Our mutual goal must be to make *The National Map* as versatile and as simple to use as it can possibly be. In short, we need to connect through improved collaboration at many levels, engage the geospatial community in the increased use of data standards, and discover the knowledge potential of *The National Map* by advancing its accessibility to a much wider audience. Trusted, interoperable geospatial data can fuel applications that help solve national challenges in new frontiers of environmental science, in finding new sources of energy, in conserving our natural heritage, and in wisely managing our natural resources.

Frederick Reuss

The Question, "What is a Map?" is More Relevant than Ever

In the era of GIS, GPS and on-the-fly, mobile mapping, when anybody with a cell phone and an Internet connection can be a cartographer, the question, 'what is a map?' is more relevant than ever. USGS topographical maps are national "urtexts" — authoritative, foundational sources of geographical knowledge about the continental United States. They have been a primary resource and reference tool for generations of geographers and cartographers, whose use and utility across a wide range of scientific disciplines and professional practices. Topographical maps have also found a wide array of non-professional uses and are distributed both in printed form and now, via *The National Map* server, over the Internet. Cartography is both a material and an imaginative undertaking. Maps have long been used both as topics and sources of inspiration by visual artists and writers. For the novelist, maps contain the seeds and threads of narratives; and mapping itself is a rich metaphor for the exploration of human geography that is storytelling. Maps can be read as more than depictions of physical terrain. They are a way of knowing the world and understanding ourselves. Beginning with the question, "what is a map?" novelist Frederick Reuss describes a unique map that was drawn and acquired under very unusual circumstances, and how it was later incorporated into his recent book, *A Geography of Secrets*. He discusses the role of maps in storytelling and narrative, and how the most subjective of maps—a map drawn from memory—combined with the most objective of maps—a 7.5' USGS topographical quadrant—can open a window into inner and outer landscapes of the past.

Panel Discussions, 11:00 a.m. – 12:00 p.m.

Surface Water Mapping Systems—NHD/NHDPlus/WBD/NED

Moderator: Jeff Simley

Panel members: Tommy Dewald, EPA; Mark Olsen, State of Minnesota; Pete Steeves, USGS; Steven Nechero, NRCS; Ricardo Lopez, City of New York; Al Rea, USGS; Karen Hanson, USGS; Dan Wickwire, BLM; Keven Roth, USGS; Brian Sanborn, USFS-NRIS

The National Hydrography Dataset has set a new standard for geospatial data sophistication that has created new possibilities for surface water science. These capabilities are continuously being expanded by improving the data in the NHD, by developing many new water characteristics derived through data integration in the NHDPlus, by detailed watershed delineation in the NHD, and by elevation derivatives processed from the NED. Powerful applications such as StreamStats and SPARROW have taken water science to new heights as a result of improved and more available geospatial data. The successes, challenges, and future development are important discussion topics to chart a course for geospatial data in surface water science.

Federal Imagery Partnerships—Status and Plans

Moderator: Russ Jackson

Panel members: George Lee, USGS; Shirley Hall, USDA; William Nellist, NGA; Kent Williams, FSA; NSGIC Representative, FSA

This session will provide an overview of the current NDOP role, activities, and plans:

- NDOP restructured to become a formal FGDC Subcommittee
- Role of USGS as the A-16 Lead Agency for Orthoimagery
- Status and plans for the National Agriculture Imagery Program
- Status and plans for the Urban Area Imagery Partnership
- Leveraging Federal Imagery Programs by State Agencies
- Community feedback

National Digital Elevation Program—Status and Plans

Moderator: Dean Gesch

Panel members: TBD

This discourse will describe the National Digital Elevation Program (NDEP) collaborative elevation acquisition efforts for the next 4 years. Led by the FEMA Riskmap program, which has the goal of updating its flood maps with improved elevation data, the USGS and other Federal and State agencies will leverage resources toward this effort to either expand the areas of interest or upgrade the quality level of the data. In addition, the following will be discussed: collaborative efforts in Alaska, a LiDAR products specification for use by all agencies, efforts at the National Geospatial Technical Operations Center (NGTOC) and Earth Research and Observation Science Center (EROS) to maximize quality control, and National Elevation Dataset (NED) and the Center for LiDAR Information Communication and Knowledge (CLICK) efforts in expectation of increasing amounts of elevation data. The NGP and NDEP will seek questions from the audience specific to the FEMA acquisition schedule, funding expectations, and the NED and USGS efforts. The forum for this presentation is not envisioned as a formal NDEP panel discussion, as many NDEP members are not expected to be present at TNM Users Conference, but may consist of some NDEP members and USGS personnel who work closely with that consortium.

National Transportation Developments

Moderator: Dick Vraga

Panel members: Andrea Johnson, Census Bureau; Marc Levine, USGS; Steve Coast, Open Street Maps; John Gottsegen, TFTN/NSGIC

This panel will discuss large-scale efforts to develop a national transportation network. An update on the "Transportation for the Nation" will be provided. Collaborative efforts from the Census Bureau, the USGS, and the U.S. Forest Service will be discussed, and the newly formed FGDC Transportation Subcommittee status will be provided. Representatives from Open Street Maps and the private sector will describe how their efforts fit into the overall national transportation framework. The audience will be asked to identify their requirements for a national approach to transportation data.

Reaching out to *The National Map's* Communities of Users

Moderator: Mike Domaratz

Panel members: Vick Lukas, USGS; Tracy Fuller, USGS; Steve Aichele, USGS

This panel will explore the development of a strategy to better engage users of *The National Map*. Participants will be encouraged to share their ideas on how the program can best identify users and document their needs for the products and services of *The National Map*.

Homeland Infrastructure, Situational Awareness, and Readiness

Moderator: Steve Hammond

Panel members: Talbot Brooks, Delta State; Laurie Jasso, USGS; Steve Alness, NGA; Tammy Barbour, DHS

The National Map is a combination of databases as well as products and services. This panel will provide a glimpse of how USGS partners are sharing robust national datasets for public domain and more sensitive official uses, leveraging existing partnerships, and actively working to increase the knowledge and use of the U.S. national grid in the public domain.

The National Map Doctor's Office , 12:00 p.m. – 3:30 p.m.

Abstracts

National Hydrology Dataset 1, 1:30 p.m. – 3:00 p.m.

WBD/NHD Integration—A New Opportunity for GIS
by Stephen Daw
U.S. Geological Survey, National Technical Operations Center, Box 25046, Bldg. 810, MS 510, Denver, CO 80225-0046, (303) 202-4418 sgdaw@usgs.gov

The National Hydrography Dataset (NHD) and Watershed Boundary Dataset (WBD) have become highly successful components of a framework for geographic information systems (GIS) in water resources. Integrating them into a common data model and linking characteristics of the two datasets will now unleash even more powerful capabilities for analysis. A highly functional data design taking advantage of geodatabase technology provides an advanced dataset that will serve sophisticated analysis for years to come while at the same time being simple and easy to use, thus giving users at all levels of expertise the ability to apply the data. Integration of the NHD and WBD presents a single data model, spatial integration, water program integration, and integration with other themes of data in *The National Map*. Developing the new dataset has demonstrated a strong interagency collaboration at all levels. This collaboration has facilitated the networking of many State and Federal agencies, which has led to a better understanding of requirements and solutions for the role of such a dataset in water science. The result today is a GIS dataset that is well on its way to providing useful new data to scientists and users for years to come. This is already apparent in national, regional, and local applications that are showing how WBD improves the management, exchange, and analysis of hydrologic data demonstrate the usefulness of the new integrated database. The usefulness of a nested hydrologic unit scheme based on natural surface conditions and combined with a water flow network cannot be underestimated for invaluable analytical and statistical purposes and applications.

Advances in Waterborne Transport Modeling Using NHDPlus
by William B. Samuels

Senior Scientist, Science Applications International Corporation, william.b.samuels@saic.com

Incident Command Tool for Drinking Water (ICWater) Protection makes use of the 1:100,000-scale National Hydrography Dataset Plus (NHDPlus), a hydrologically connected river network that contains over 3 million reach segments in the United States. This allows for both downstream and upstream river network navigation. Mean flow and velocity have been calculated by the U.S. Environmental Protection Agency (USEPA) for each reach segment. The mean flow and velocity data are scaled by real-time flow from nearby gaging stations. A new version of ICWater (version 3.0) has been developed that includes the ability to model runoff from land applications or atmospheric deposition of contaminants. The runoff calculations make use of the NHDPlus catchments (and associated land use attributes), elevation grids, hydrologic soil groups, and National Oceanic and Atmospheric Administration (NOAA) Quantitative Precipitation Forecasts. The transport of contaminants from atmospheric deposition in catchments is based on the following methodology and steps: (1) Deposition areas within each NHD catchment are determined by intersecting the deposition model output polygons with the catchments layer, and a grid is established for each deposition area within catchments. (2) From each grid cell in a deposition area, a total travel time from each contaminated cell to the catchment outlet is determined assuming that only shallow concentrated flow occurs for the entire transport route. (3) The total time of travel from each contaminated cell to the catchment outlet is determined by accumulating travel times from cell to cell along the flow route. By calculating the travel time that contamination takes to move from each grid cell individually within a deposition area, the method mimics a routing-like approach to transport the total accumulated mass in each area. In the approach taken, rainfall is assumed to dissolve and carry the entire mass deposited from each cell in the contaminated area along the flow path to the outlet. The mass of one cell is then introduced at the outlet with the travel time being the lag time between the occurrence of rainfall and the injection at the outlet. In addition to the surface runoff calculations, the NOAA tide gage locations have been incorporated into the system to designate when the downstream trace has entered a region under tidal influence. A boundary condition source term can be generated as input to a two-dimensional estuarine model. The entire system is built upon the ESRI ArcEngine component of the Commercial Joint Mapping Toolkit (CJMTK).

Watershed Boundary Dataset (WBD) Applications
by Karen M. Hanson,[1] Brian W. Sanborn,[2] David R. Brower,[3] Dan Wickwire,[4] and Tommy Dewald[5]

[1]*U.S. Geological Survey, Watershed Boundary Dataset Product Manager, 2329 W. Orton Circle, West Valley City, UT 84119, khanson@usgs.gov*
[2]*U.S. Forest Service, Natural Resource Information System, WO-EMC 4077 Research Way, Corvallis, OR 97333, (541) 750-7151 bsanborn@fs.fed.us*
[3]*Cartographer/GIS Specialist, USDA Natural Resources Conservation Service, Washington State Office, 316 West Boone Avenue, Suite 450, Spokane, WA 99201, (509) 323-2962 dave.brower@wa.usda.gov*
[4]*Bureau of Land Management, Portland, OR, (503) 808-6272 dwickwir@or.blm.gov*
[5]*USEPA Office of Water, 1200 Pennsylvania Ave. NW (4503T), Washington, DC, (202) 566-1178 Dewald.Tommy@epa.gov*

The Watershed Boundary Dataset (WBD), a component of *The National Map*, is a nationally consistent, seamless, and hierarchical hydrologic unit dataset based on topographic and hydrologic features across the United States. This dataset provides a consistent framework for local, regional, and national applications to manage, archive, exchange, and analyze data by hydrologic features. The usefulness of hydrologic units in a variety of sizes based on natural surface-water flow and topography cannot be overestimated in hydrologic analytical and statistical applications. During this session a panel will discuss a multi-agency range of applications for these data, including watershed management, water-quality initiatives, watershed modeling, resource inventory and assessment, fire assessment and management, and Total Maximum Daily Load. The use of partnerships, a robust production program, thorough data structure, and ease of access are important factors in the success of this aspect of *The National Map*.

The National Map Data Themes – Elevation, 1:30 p.m. – 3:00 p.m.

Finding the Streams in the Coastal Plain—Will LiDAR Help?
by Silvia Terziotti

U.S. Geological Survey, North Carolina Water Science Center, 3916 Sunset Ridge Rd., Raleigh, NC 27607, seterzio@usgs.gov

The flat terrain of North Carolina's Coastal Plain has proven to be a challenge when mapping watersheds and stream networks. Furthermore, the Coastal Plain's stream network has been extensively altered with channels, ditches, and drainage canals, which add additional complexity to watershed mapping. The 1:24,000-scale high-resolution National Hydrography Dataset was mapped with remarkable accuracy and detail. However, as the National Elevation Dataset is enhanced with Light Detection and Ranging (LiDAR)-derived elevation data, the need for a local-resolution stream network that is accurately aligned with the elevation surface becomes more important, especially for hydrologic applications such as watershed modeling. Examination of data from one county in the North Carolina Coastal Plain, using LiDAR collected at two different times and resolutions, is used to illustrate methods of automated stream and watershed delineation. Examples of products that can be extracted in areas of very low relief and extensive artificial drainage are compared on the basis of the source LiDAR dataset. Issues such as post spacing, breakline positioning, and documentation of datasets are important elements in stream and watershed mapping.

Status of Alaska Orthoimagery and Elevation Mapping and Alaska Statewide Mapping Program Overview
by Tom Heinrichs[1] and Dayne Broderson[2]

[1]*University of Alaska Fairbanks - Geographic Information Network of Alaska, Director, tom.heinrichs@alaska.edu*
[2]*University of Alaska Fairbanks - Geographic Information Network of Alaska, dayne@gina.alaska.edu*

Alaska currently has the oldest and least accurate maps of any State in the United States. Alaska is updating its statewide orthoimagery and elevation data for the first time in more than 50 years. New contracts for statewide orthoimagery and digital elevation model (DEM) collection are under way. A total of 603,000 square kilometers (39 percent of the State) of new orthomosaic source imagery was collected in 2009 and 2010. The orthoimagery project is fully funded for statewide completion by 2014. In 2010, 157,000 square kilometers (10 percent of the State) of new airborne interferometric synthetic aperature radar (IfSAR) DEM data were collected. Funding for statewide DEM completion is being sought. Both projects are joint State-Federal partnerships. Currently, there exists no statewide orthoimage base for Alaska. Unrectified photography from the 1950s U.S. Geological Survey (USGS) statewide mapping campaign and unrectified photography from the 1978-86 Alaska High-Altitude Photography project exist, but no statewide digital orthoimagery layer other than Landsat is available for Alaska. Most populated areas, including cities and remote villages, and some areas with economic resources, have been imaged at high resolution and have accurate orthoimagery available; these represent less than 15 percent of the State. The current National Elevation Database (NED) for Alaska is coarse resolution: 2-arcsecond postings, roughly 30x60-meter cells at Alaska's high latitude. It was created by digitalizing the USGS topographic maps of 1950s vintage, many of which have significant accuracy limitations. The NED is known to contain large errors in some areas of the State. Less than 10 percent of the State currently has higher accuracy DEMs created with Light Detection and Ranging (LiDAR), airborne IfSAR, or photogrammetry. The Alaska Statewide Digital Mapping Initiative, a multi-agency partnership, is addressing these shortcomings through two projects. One is the creation of a new statewide orthomosaic imagery base layer at 1:24,000 National Map Accuracy Standards (NMAS) accuracy (12.2-meters CE90). The entire State (1.56 million square kilometers) will be imaged with the SPOT 5 satellite, and a 2.5-meter spatial resolution, multi-spectral, pan-sharpened orthoimage will be produced by 2014. Source imagery collection is 38 percent complete as of 2010. The second project is the collection of an improved-accuracy DEM. The first phase is under way with 157,000 square kilometers (10 percent of the State) of airborne IfSAR data collected in 2010. A 5-meter post spacing, 20-foot contour interval accuracy equivalent (3-meter LE90) DEM, and radar backscatter intensity image will be produced. Planning for orthoimagery refresh and completion of the statewide DEM are under way.

LiDAR Acquisition of the Atchafalaya River Basin, Louisiana
by Jochen A. Floesser

GISP, PMP Program Manager, Civil Works – 3001 International Integrated Intelligence Systems, Northrop Grumman Information Systems, 301 Voyager Way NW, Huntsville, AL 35806, (256)-830-3435 jochen.floesser@ngc.com

The Atchafalaya River Basin in south-central Louisiana functions as a regulated distributary system to the Mississippi River. Both natural changes and engineering projects have had impacts on the basin's flow system, resulting in segmented drainage regimes that are less connected to the river and that resulted in decreased water quality. In order to better evaluate management plans, the Louisiana Department of Natural Resources (LDNR) Atchafalaya Basin program is backing the development of the Natural Resources Inventory and Assessment System (NRIAS). This system incorporates and manages a variety of geospatial information and scientific data. Among other inputs, high-resolution elevation data are required in order to determine the basin's flow patterns. Northrop Grumman was tasked by the U.S. Geological Survey (USGS) to perform light detection and ranging (LiDAR) data collection during low water stages and leaf-off conditions for the 980 square mile Atchafalaya River Basin project area in late fall of 2010. The presentation outlines the work involved during planning, aerial acquisition, ground control survey, calibration, and processing of the LiDAR data at nominally 1-meter point spacing, as well as the generation of the resulting elevation data products. In preparation for LiDAR acquisition, both the vegetative status and the water levels were monitored in the Area of Interest (AOI) during fall of 2010 with the assistance of the U.S. Army Corps of Engineers' (USACE) Engineer Research and Development Center (ERDS) on-site staff. Field observations and data analysis from a preliminary LiDAR test flight in late November showed that the canopy cover in the area was sufficiently thinned to allow for the capture of LiDAR data in accordance with the statement of work specifications. In early December, aerial capture of LiDAR data and control survey were accomplished during the desired environmental conditions (low water stages, leaf-off). During the following weeks, the acquired airborne global positioning system (GPS) and inertial measurement unit (IMU) data were processed, and LiDAR data were calibrated (boresighted). The data were then cut into tiles and underwent classification and hydro-breaklining. The final project deliverables include the raw point cloud data (.las), classified point cloud (.las), bare earth surface (raster DEM), breaklines, survey control data, metadata and auxiliary information, and various project and production reports. The data products will ultimately be used as basic site plan information and elevation mapping for hydraulic modeling, design, environmental management, and engineering studies covering pertinent issues such as inundation impacts and flow control studies. Results from these studies will provide insight into predicting landscape change, promoting restoration of ecosystems, and mitigating risks associated with anthropomorphic and natural hazards.

The National Map Partnerships 1, 1:30 p.m. – 3:00 p.m.

What Motivates Partnerships—Greed, Altruism or Pragmatism? And Where Do We Go from Here?
by Jay B. Parrish

Professor of Practice, Dutton Institute, Penn State University, jbp3@psu.edu

The National Map has been predicated upon an assumption of partnerships between governments at all levels. But what motivates those partnerships? Some common motivators would include: (1) Loyalty—a desire to maintain and extend relationships forged through liaison contacts. (2) Greed—the desire to get more than received. (3) Desire to cooperate—not altruism so much as a contractual agreement. (4) Pragmatism—joining with a Federal program results in a mutually beneficial expansion of funding. (5) Altruism—because it is the right thing to do. (6) Politics—this can be both at the Federal level or even with respect to trade and professional organizations-- an emphasis on policies which may or may not favor partnership. Various partnership models have been used at both the Federal and State levels. The U.S. Geological Survey has provided matching funding to great effect. This partnership model has appealed to all of the above motivators, with sufficient funds so as to avoid a backlash greed factor as in the Wolfenschiesse case. The motivator in each State (or county) will differ with circumstances. The question is how to find the proper motivator for the diverse situations found in 50 States while retaining some semblance of a national program. Contrary to common sense, one solution is to create a top-down approach with federally devised rules and rewards. This may be successful because the rules have changed. Bad economic times have disempowered States and local governments to a greater extent, leaving the Federal Government with more leverage. Given the difficulties inherent in integrating data from different localities, the need for a Federal approach and the emergence of a private industry as a source for extensive aerial imagery, it is an opportune time for a rethinking of what a partnership is and what motivates partners.

New Jersey's Ongoing Partnership with *The National Map*
by Seth Hackman,[1] Dave Anderson,[2] and Roger Barlow[3]

[1]New Jersey Department of Environmental Protection, GIS Specialist, seth.hackman@dep.state.nj.us
[2]U.S. Geological Survey, 1400 Independence Road, MS 647, Rolla, MO 65401, danderson@usgs.gov
[3]U.S. Geological Survey, 12201 Sunrise Valley Drive, MS 433, Reston,VA 20192, rbarlow@usgs.gov

The State of New Jersey, through its partnership with U.S. Geological Survey (USGS), has embarked on three projects to enhance statewide data and *The National Map*. This presentation will highlight the major points and details of these efforts. (1) National Hydrography Dataset (NHD): The New Jersey Department of Environmental Protection (NJDEP) has signed a stewardship agreement with the USGS to update and maintain the NHD for the State. Since that agreement was signed, NJDEP has effectively conflated its 1:24,000-scale high-resolution hydrography to 1:2,400 local-resolution scale data statewide. USGS has recognized this conflation effort to be the first time any State has loaded 1:2,400 scale or larger into *The National Map*. (2) Watershed Boundary Dataset (WBD): NJDEP also acts as stewards of the WBD for the State. New Jersey has recently updated 6th level subwatersheds to match the new local-resolution NHD. Moreover, NJDEP personnel routinely use 7th level subwatersheds to submit integrated report data in conjunction with the Federal Clean Water Act. NJDEP is currently undertaking a project to use digital elevation models derived from varying levels of Light Detection and Ranging (LiDAR) imaging to update those 7th level boundaries for the WBD. (3) Geographic Names Information System (GNIS) / Names: Through various state-wide feature updates that have taken place over the previous 10-15 years, NJDEP is prepared to update certain feature classes of the GNIS as part of an effort to improve the US Topo project for New Jersey. Additionally, NJDEP plans on submitting hundreds of previously undocumented hydrographic feature names from local sources to the GNIS. These names were discovered as part of the NHD conflation project and will make streams, thoroughfares, and other natural feature classes in the GNIS significantly more robust.

U.S. Forest Service/U.S. Geological Survey Transportation Data Sharing Initiative
by Karen Nabity,[1] Chuck Matthys,[2] and Greg Matthews[3]

[1]U.S. Forest Service, Geospatial Service and Technology Center, Group Leader - Data Services and Products, knabity@fs.fed.us
[2]U.S. Geological Survey, National Geospatial Technical Operations Center, Box 25046, Denver Federal Center,
MS 510, Denver, CO 80225, (303) 202-4447 cpmatthys@usgs.gov
[3]U.S. Geological Survey, National Geospatial Technical Operations Center, Box 25046, Denver Federal Center,
MS 510, Denver, CO 80225, (303) 202-4446 gdmatthews@usgs.gov

Over the past two and half years, the U.S. Forest Service (FS) and the U.S. Geological Survey (USGS) have embarked on a partnership to create an improved roads dataset for the Nation. So far, this collaboration has resulted in the integration of FSTopo roads into *The National Map* transportation database and USGS topographic mapping products covering FS lands for 28 States. This presentation will discuss the background and purpose for starting this program and the resulting joint agency cooperation to develop methods and software programs to integrate the FS roads data. Additionally, future collaboration plans for data integration will be presented.

The National Map Research, 1:30 p.m. – 3:00 p.m.

Semantic Web Technology for *The National Map*
by Dalia Varanka

Research Geographer, Center of Excellence for Geospatial Information Science (CEGIS), dvaranka@usgs.gov

The Semantic Web is composed of techniques for automatically linking data across the Internet. To provide increased access and utility to *The National Map* for Semantic Web users, sample data were converted from ESRI shapefiles and geodatabases to N3 triples, a format that is easily used for semantic technology. Triples consist of two nodes and an arc relating the two nodes together. The semantic meaning of the data is specified by a universal resource identifier (URI), and when combined as triples, takes the form of a subject-predicate object. The subjects and objects of triples can be drawn from the Best Practices data model, and the predicate is drawn from the Open Geospatial Consortium (OGC) standard topological spatial relations. Logic axioms are applied using Resource Description Framework (RDF) and Web Ontology Language (OWL) to form the ontology that controls the linkages between the N3 triple data. The converted sample data can be queried using an endpoint for the SPARQL Protocol and RDF Query Language of the Semantic Web. The endpoint to query the data is linked on the Building a Semantic Technology for *The National Map* Research Project's home page (http://cegis.usgs.gov/ontology.html). These results build toward a vision of topographic data in the form of ontology design patterns (ODP), which are assemblages of the component parts of complex features. ODP can be made available at an ontology repository for open access and public reuse. User cases are described to explore new operations of semantic topographic data to expand the potential for *The National Map* applications.

QUAD-G: New Technology for Automated Georeferencing of Scanned Quadrangles
by James E. Burt,[1] Jeremy White,[2] Greg Allord,[3] and A-Xing Zhu[4]

[1]*University of Wisconsin-Madison, Professor of Geography, jeburt@wisc.edu*
[2]*University of Wisconsin-Madison, jeremy@blueshirt.com*
[3]*U.S. Geological Survey, 465 Science Drive, Suite A, Madison, WI 53711, gjallord@usgs.gov*
[4]*University of Wisconsin-Madison, azhu@wisc.edu*

This paper describes QUAD-G, a new software tool for georeferencing scanned topographic quadrangles without operator intervention. Developed as part of the U.S. Geological Survey Historical Quadrangles Scanning Project, the program is distributed as a Windows executable and as open-source C# files (see http://geography.wisc.edu/research/projects/QUAD-G). Here we present technical details of QUAD-G along with performance measures accumulated over 6 months of use in the Historical Quadrangle Scanning Project (HQSP). In contrast to existing programs that require on-screen digitizing of control points and manual population of dialog boxes, QUAD-G uses pattern matching to automatically identify control points and acquires map sheet information from a metadata file. Least squares with cross-validation is used to establish and diagnose a best-fit second degree polynomial transformation between image and geographic coordinate systems. Substandard fits are flagged according to a user-specified error tolerance. These arise if the pattern search fails to find an internally consistent suite of control marks in the image. Transformations passing the threshold are applied to the scan, resulting in a new GeoTiff image in geographic coordinates. The output image preserves the ground resolution found in the original. In addition to reporting on model fitting errors, QUAD-G performs a quality analysis on the output image. In particular, the image is searched for control marks and deviations between search results, and known positions are noted. These are stored as both tabular data and visual displays for later review. Requiring only a few minutes of operator setup, the program is typically run in unattended mode on batches of hundreds of maps as part of the HQSP. Each scan is required to have a cross-validated RMSE of less than 6 pixels or it will not be georeferenced. Of the approximately 100,000 scans processed to date, about 80 percent passed this test and were successfully georeferenced with no supervision. Quality analysis of output files confirms the low model errors—control marks are typically located within a few meters of expected locations. As we show in the paper, scans not passing the threshold are usually from old maps whose line quality is too degraded to be confidently detected by pattern matching. As we also demonstrate, addressing these with QUAD-G in its manual mode is much faster than using a standard tool.

Establishing Classification and Hierarchy in Populated Place Labeling for Multiscale Mapping for
The National Map
by Stephen J. Butzler, Cynthia A. Brewer, James E. Thatcher, and Wesley J. Stroh

*Department of Geography, Pennsylvania State University, 302 Walker Building, University Park, PA 16802, sjb204@psu.edu,
cbrewer@psu.edu, jethatcher@gmail.com, wjstroh@psu.edu*

Current Federal mapping standards call for the use of the Geographic Names Information System (GNIS) point layer for place-ment of U.S. populated place labels. The point layer contains limited classification information and therefore constrains the balanced labeling of place. In turn, this affects map quality for database-driven, multi-scale reference mapping, such as maps served by *The National Map* Viewer from the U.S. Geological Survey (USGS). Database-driven mapping often relies simply on what labels best fit in the map frame. Our research investigates alternative sources for populated place classification in order to improve mapping practices so that populated place labels accurately reflect the relative importance between actual places. We examined several populated place polygon categorizations defined by the U.S. Census Bureau, such as incorporated place, census designated place (CDP), and economic place. Within each of these polygon layers we investigate relevant attributes from the decennial and economic census such as population for incorporated place and CDPs, and the number of employees for economic places. The data selected are available for the entire country to serve national mapping requirements. We differentiate a fourth category of GNIS populated place points, essentially neighborhoods and related features, which are neither incorporated places, CDPs, nor economic places. Populated places in this fourth class do not have federally defined boundaries, necessitating an alternative method for determining hierarchy in label presentation through scale. Work presented at the AutoCarto 2010 con-ference on this theme examined Pennsylvania and Colorado as test cases. This talk extends the place names work to the entire United States. In addition, we compare our initial maps with place name hierarchies produced from a combination of GNIS with economic and decennial census attributes to maps produced using a simpler data model that augments GNIS only with decennial census attributes.

The National Map and Emergency Operations, 1:30 p.m. – 3:00 p.m.

The Nationally Consistent Source of Base Mapping, *The National Map* and US Topo: A Sample of The End User's View and Requirements
by Major William J. Schouviller[1] and Neri G. Terry, Jr.[2]

*[1]USMC HQMC IPI, (703) 695-3118 william.j.schouviller@nga.mil
[2]HQMC IPI, neri.terry@usmc.mil*

Whether the situation is a cross-jurisdictional emergency response or Boy Scouts going to the field, there is a need for a nation-ally consistent base map provided by the U.S. Geological Survey. The base (topographic) map of the U.S. begun by Maj. John Wesley Powell continues to be a critical national dataset. One of its great attributes is that it is readily available to unrestricted use and enhancement by any and all users as a public domain dataset. Today high-end users require both individual vector and imagery datasets from *The National Map*, while geographic information system (GIS) disadvantaged users require the traditional cell-based finished base map, albeit today in both a digital as well as hardcopy format. Because the vast majority of users are GIS disadvantaged, there remains the need to maintain an up-to-date set of finished base maps such as US Topo, drawn from *The National Map* data. This presentation will discuss some of the needs and requirements in both *The National Map* and US Topo. Much great work has been accomplished in building out *The National Map* and the extraordinary production we have witnessed in US Topo products, but there remains the need for improvements that will be addressed.

Demonstration: Using *The National Map* as a Common Operating Picture Viewer for Visualization of National Critical Infrastructure and Key Resources
by Richard A. Benjamin

National Geospatial-Intelligence Agency (NGA) Office of the Americas, Domestic Preparedness Branch, Richard.A.Benjamin@nga.mil

The Homeland Infrastructure Foundation-Level Data (HIFLD) Working Group conducted an assessment of Infrastructure Geospatial Viewers. The assessment produced the Infrastructure Information Community Model (IICM) that is a graphical representation depicting the data, product processes, and key infrastructure viewers that are utilized by Federal, State, local, and public mission partners. The assessment documented the capabilities of various infrastructure viewers including the technical and operational interactions. The new version of Homeland Security Infrastructure Program (HSIP) Gold is scheduled to be released in 2011 and consists of 450 (increased by 107) vector layers composed of commercial, Federal- and State-level datasets. The USGS' *The National Map* contributes important base layers to this comprehensive data holding that is used by many organizations. A demonstration of the *The National Map* visualization tools and Web services specific to *The National Map* Viewer, Palanterra X3, and iCAV Next Generation will be performed as examples of some government solutions available to the Homeland Defense and Homeland Security community. Preference of systems depends on the user. For example, what may work in the disaster community may not work in the law enforcement community. Regardless of infrastructure viewer, primary focus moving forward is on the geospatial data and whether or not they can be ingested into the viewer of choice via Web services (that is, ArcGIS Services, Web-Mapping Services, KMLs, and GeoRSS feeds). The demonstration will include a walkthrough of how users may add Web services to *The National Map* to highlight the benefits of the capability, as well as some of the challenges and needs. Web services that are developed following established standards and include a reference implementation or implementation profile provide immediate seamless accessibility. However, most Web services are not developed in this manner. Web services are especially valuable to access dynamic data. Although Geospatial Web Service Standards are progressing toward an enterprise solution, greater specificity is needed to reduce ambiguity and improve interoperability.

The National Map and the National Guard: New Synergies for Domestic Operations
by Brian Cullis

Critigen, GeoGuard Lead Consultant, 90 South Cascade, Suite 700, Colorado Springs, CO 80903, (719) 477-4968, Brian.Cullis@CH2M.com

In the fall of 2010, the National Guard Bureau (NGB) and the U.S. Geological Survey (USGS) formally agreed to collaborate to acquire, maintain, archive, and provide geospatial and remote sensing information, data, and other services to better conduct Defense Support of Civilian Authorities (DSCA) and support the emergency management community. Concurrently, the NGB endorsed an initiative called GeoGuard which aims to organize and mobilize a new National Guard geospatial enterprise to better leverage GIS for all National Guard missions. This paper will describe the current and potential synergies being realized for the DSCA mission by the National Guard through teaming with the USGS and leveraging of the USGS liaison network and products and services. Specific attention will be given to the role of Wideband Imagery Dissemination System–Broadcast Request Imagery Technology Environment (WIDS-BRITE) Unclassified Domestic Imagery Dissemination Manager (UDIM), Hazard Data Delivery System (HDDS), US Topo, and *The National Map* in delivering shared situational awareness for the DSCA mission.

The National Map Data Themes – Elevation, 1:30 p.m. – 3:00 p.m.

National Elevation Dataset, Applications of *The National Map*
by Sandra Poppenga

U.S. Geological Survey, Earth Resources Observation and Science (EROS) Center, EROS Center, 47914 252nd Street, Sioux Falls, SD 57198, spoppenga@usgs.gov

The National Elevation Dataset (NED) is the primary elevation data product produced and distributed by the U.S. Geological Survey (USGS). The NED serves as the elevation layer of *The National Map* and provides basic elevation information for earth science studies and mapping applications in the United States. Scientists and resource managers use NED data for global change research, hydrologic and hydrographic modeling, resource monitoring, mapping and visualization, and many other applications. The NED first became available in 1999 at a 30-meter resolution for the conterminous United States. Since 2003, the NED has evolved into a multi-resolution dataset that not only includes 1-arc-second (~30-meter) and 1/3-arc-second (~10-meter), but also 1/9-arc-second (~3-meter or better) elevation data that are derived mostly from light detection and ranging (LiDAR) 3-D point cloud data. The increasing use of LiDAR 3-D point cloud data to generate digital elevation models (DEMs) improves the overall vertical accuracy, or vertical component, of the NED, the elevation layer of *The National Map*, thus fulfilling the goal of the USGS to provide the best available elevation data. With increased vertical accuracy and horizontal resolution, the elevation-based applications of *The National Map* have abounded, especially in research studies of climate change effects and depletion of natural resources. The USGS Earth Resources Observation and Science Center (EROS) Topographic Science Team utilizes the NED for studying coastal vulnerability to sea level rise and storm surge modeling to identify populations at risk from natural disasters. Topographic/bathymetric data-integration research strives to provide better spatial accuracy of coastlines, which will enhance our understanding of the landscapes subject to inundation. The improved vertical accuracy and horizontal resolution of the NED allow more accurate and detailed contours, hydrologic and hydrographic information to support flood mitigation efforts, and identification of locales vulnerable to potential disasters. These elevation applications of *The National Map* emphasize the importance of data integration and leveraging of evolving technologies to meet national priorities and global trends.

LiDAR Quality Assurance Challenges and Solutions
by Hans Karl Heidemann

U.S. Geological Survey, EROS, 47914 252nd Street, MS SAB, Sioux Falls, SD 57198, kheidemann@usgs.gov

With the introduction of the U.S. Geological Survey (USGS)-National Geospatial Program (NGP) Light Detection and Ranging (LiDAR) Guidelines and Base Specification (Spec), the USGS and the LiDAR community at large are now taking a broader view of LiDAR as a data type, specifically by focusing more attention on the source point cloud data as opposed to the derived bare-earth digital elevation model (DEM). Quality assurance (QA) of the point cloud data has become a new focus and the topic of much activity throughout the industry. Only very recently has development begun on tools and procedures to effectively perform QA of the source point as required by the Spec. Similarly, the American Society for Photogrammetry and Remote Sensing (ASPRS) has begun efforts to establish standard industry practices for the assessment and reporting of LiDAR data calibration accuracy. These efforts signal a significant and long-needed paradigm shift in the LiDAR industry, moving away from an almost exclusive concentration on derived DEMs to the recognition of the point cloud as the primary data and source for numerous other derivatives. Assurance of the integrity of the source point cloud data will allow LiDAR technology to serve a broader range of applications and enable more reliable integration of individual collections for wider-area analysis. This presentation will survey some of the more troublesome aspects of LiDAR data QA, present some of the tools currently available or in development to assist in LiDAR QA, and provide an update on the ASPRS activities on LiDAR calibration standards.

High Resolution LiDAR in Ecological Research in the Pacific Northwest
by Patricia Haggerty

U.S. Geological Survey, Forest and Rangeland Ecosystem Science Center, 3200 SW. Jefferson Way, Corvallis, OR 97331, phaggerty@usgs.gov

The Forest and Rangeland Ecosystem Science Center (FRESC) aims to be a keystone provider of scientific knowledge to land management agencies, with emphasis on agencies within the Department of the Interior. In order to enhance our capabilities to provide a scientific foundation for conservation, restoration, and mitigation programs at multiple spatial scales, we are increasingly using very high resolution light detection and ranging (LiDAR) data obtained through the Oregon and Puget Sound LiDAR consortiums. Airborne LiDAR projects acquired through the Oregon LiDAR Consortium average 8 pulses per meter or better and can have a vertical resolution of 0.03 meter. However, even at these relatively high resolutions, capturing an accurate ground elevation model can be difficult. Forests in our region may reach 91 meters (300 feet) in height and consist of complex, multiple tiers of conifers and deciduous hardwoods with thickly vegetated ground cover obscuring a complex geomorphology that is characterized by earth flows and landslides. The resolution of traditional elevation datasets is challenged in this region, yet in spite of the canopy obstructions, high resolution LiDAR has redefined the expectations of digital elevation models. Examples of FRESC projects which include LiDAR data include wildlife habitat modeling in old growth and secondary forests managed by the Bureau of Land Management (BLM), geomorphology of mountain headwater streams in Mount Rainier National Park, forecasting the potential spread of invasive species following the decommissioning of two dams in the Elwha River watershed, Olympic National Park, and the potential effects of sea level rise on shorebird use of rocky intertidal zones on the Pacific Coast. The BLM in Oregon is seeking to develop two silvicultural laser laboratories to explore the utility of very high resolution LiDAR and is supporting research investigating a number of remote sensing techniques that can efficiently describe a forested ecosystem's topography, habitats, standing biomass, and soil carbon stocks.

Board of Geographic Names – Domestic Names Committee, 1:30 p.m. – 4:00 p.m.

Legislative Approaches in the Geospatial Community 4:00 p.m. – 5:00 p.m.

Moderator: Chris Trent

A small round table of communications specialists will offer a brief introduction to their work and share their perspectives on the intersection of geospatial information with legislative affairs. Discussion will follow on hot topics and questions and comments from attendees.

National Hydrography Dataset 2, 3:30 p.m. – 5:00 p.m.

Colorado's Insights to Stewardship of the NHD
by Chris Brown

GIS Program Manager, Colorado Division of Water Resources, chris.k.brown@state.co.us

The Colorado Division of Water Resources (DWR) has been actively involved in Stewardship of the National Hydrography Dataset (NHD) for nearly 3 years. During this time the Colorado DWR has reviewed the data for the entire State, making corrections and taking notes to help develop strategies for future maintenance of the NHD and use of the NHD by the State of Colorado. As a result of this process, the DWR has identified three issues important to the success of the NHD in the State: Tracking and Submittal of New Geographic Names Information System (GNIS) Names, Processes for Addition of Event Data to the NHD, and Targeted Photo Revision. Tracking and Submittal of New GNIS Names will cover Colorado's process of tracking GNIS errors and the submittal of new names to the Board of Geographic Names. The types of data that need to be tracked for submittals and the importance of the GNIS to Colorado will be discussed. DWR personnel have recently started to add events (dams, streamgages, and diversions) to the NHD and have worked out a process for event inclusion. Types of events that are important to Colorado, the process DWR uses, and the focus of future event additions will be discussed. Targeted Photo Revision will cover Colorado's suggested approach to building an NHD that not only is a good cartographic representation of the hydrographic data but also is useful to modeling. Selection of candidate areas to be photo revised will be discussed, as well as the proposed changes to the cartographic representation of certain features.

National Hydrography Dataset Web Edit Tool (NHD-WET)
by Phillip Henderson

GIS Manager, Alabama Department of Economic and Community Affairs, phillip.henderson@adeca.alabama.gov

The NHD Web Edit Tool, or NHD-WET, is a Web-based editing tool designed to enhance the functionality that is already provided by the existing National Hydrography Dataset (NHD) GeoEdit Tool. The NHD-WET makes it possible for State Stewards to receive input from local partners who have a vast knowledge of their hydrologic network. These partners increase the number of users making corrections to the datasets, which will result in a higher quality dataset on a much faster timetable. The missing ingredient in tapping into this valuable resource is the availability and the ease of use that the NHD-WET provides. Access to NHD-WET can be easily achieved by using Web applications that can run on a Web browser. Since NHD-WET is Web based, it will not require additional desktop client license costs or fees and can be easily accessed by a Web browser by the local stewards. The NHD-WET has two user levels. The first level is an entry-level Web attribute editor tool. This editor flags errors for review by the State Steward. The second level is a vector and attributes editor for submitting corrections to the State Steward. By augmenting the existing NHD Geo Edit Tool, more users are able to participate in the process and aid the State Steward in expediting the revision of the NHD datasets.

Applications of the National Hydrography Dataset
by Kathy Isham

U.S. Geological Survey, Box 25046, Denver Federal Center, MS 510, Denver, CO 80225 krisham@usgs.gov

The National Hydrography Dataset (NHD) is the surface-water component of *The National Map*. The NHD is a digital vector dataset that provides a framework for analysis by using geographic information systems. The NHD is used by many organizations not only for general purpose mapping but also for the analysis of surface water information. The NHD benefits many users working in hydrology, pollution control, fisheries biology, and resource management. A variety of applications in these fields is presented to explore the use of the NHD in geographic and scientific analysis by providing examples of the data, tools, and maps that make the analysis possible.

The National Map Data Themes – Historic Maps, 3:30 p.m. – 5:00 p.m.

Scanning and Georeferencing Historical USGS Quadrangles
by Greg Allord

U.S. Geological Survey, 465 Science Drive, Suite A, Madison, WI 53711, (608) 204-0082 ext. 250 gjallord@usgs.gov

The U.S. Geological Survey (USGS) Historical Quadrangle Scanning Project (HQSP) is scanning all scales and all editions of approximately 250,000 topographic maps published by the USGS since the inception of the topographic mapping program in 1884. This scanning will provide a comprehensive digital repository of USGS topographic maps, available to the public at no cost. When physical and cultural features change over time, maps are updated, revised and new editions printed. Although they are out of date, these historic maps are often useful to scientists, historians, environmentalists, genealogists, and others researching a particular geographic location or area. A series of maps of the same area published over a period of time can show how some areas looked as early as 1884, before current development, and provide a detailed view of changes over time. Historical maps are stored in a limited number of collections and are not readily available to the public. The USGS has begun to convert these historical printed topographic quadrangles to an electronic format. This project serves the dual purpose of creating a master catalog and digital archive copies of the irreplaceable collection of topographic maps in the USGS Reston Map Library as well as making the maps available for viewing and downloading. This presentation will describe how the HQSP is accurately cataloging and creating metadata to accompany high-resolution, georeferenced digital files representing the lithographic maps. Each map image is scanned as is and captures the current content and condition of each map sheet. The project provides ready access to maps that are either no longer available for distribution in print or are being replaced by the new generation of US Topo maps. Georeferencing of the map files, that is, tying them to a known earth coordinate system, enables them to be imported into Geographic Information Systems so that they can be overlain with other geospatial (map) data from other sources such as from *The National Map*. The potential for research that analyzes change over time is becoming increasingly recognized by the geospatial community, and this project will provide published lithographic USGS maps in georeferenced digital formats. With georeferencing, the historical maps can be combined with current data from *The National Map*. The product will be delivered in a compressed GeoPDF format and as GeoTIFF images with embedded metadata.

The National Map Data Themes – Structures, 3:30 p.m. – 5:00 p.m.

Geospatial Collaboration Tools Aid Louisiana Structures Project
by Craig Johnson[1] and Chris Cretini[2]

[1]Director, Louisiana Geographic Information Center (LAGIC) at Louisiana State University, cjohnson@lsu.edu
[2]U.S. Geological Survey, NSDI Partnership Office, 700 Cajundome Blvd., Lafayette, LA 70506,
(337) 266-8621 cretinic@usgs.gov

The Louisiana Geographic Information Center (LAGIC) has been the recipient of two U.S. Geological Survey (USGS) Cooperative Agreements for building a Structures dataset for Louisiana Coastal Parishes (counties). The project has a number of partners. The four Regional Planning District offices that are located within the Louisiana Coastal Zone have provided the initial contacts with local governments. Local universities have provided technical advice and outreach to local community leaders. The USGS and the Louisiana Office of Homeland Security and Emergency Preparedness have provided high-resolution imagery of the coastal parishes to help in identifying structure locations. The Louisiana Department of Transportation has provided highway and other datasets. In Phase I, four parishes in Southeast Louisiana were selected (months prior to the British Petroleum (BP) Oil Spill). These same parishes were heavily impacted by the spill. The spill reinforced the importance of structure data as an input to geospatial analysis. Many of the companies and crews that came to assist during the cleanup needed to know the location of local government offices and emergency preparedness facilities. Phase II is now under way with four additional parishes contributing the parcel data, addresses, and structure information needed to build this critical dataset. Although not part of the original contract, the LAGIC has updated the Geographic Names Information System (GNIS) datasets as we have updated the structures datasets and have provided these data to State and Federal agencies. We have used the Department of Homeland Security data model for organizing attribute data, although not all of the data that could potentially be categorized within that data model have been completed. Two Parishes that were not originally included, because they did not require technical assistance, have voluntarily contributed data to the project, allowing for the display of structure data for larger areas of the coast than were originally envisioned. Outreach to the local government component will be essential to ensuring that the data are maintained to USGS standards over time. This work has been aided by the use of geographic information system (GIS) collaboration tools that allow local governments to review the data online and make changes to the data in real time. We have been using the collaborative capabilities built in to ArcGIS 10 software. This is critical to those parishes that do not have GIS software or GIS editing capability, yet are the custodians of these critical datasets. Local governments are the only organizations that can effectively maintain these datasets over time.

The National Map Data Themes – Parcels, 3:30 p.m. – 5:00 p.m.

SEAMLESS USA, Realities and Prospects of a 3,140 County National Parcel Layer
by Dennis H. Klein

President and CEO, Boundary Solutions, Inc., 240 Miller Avenue, Mill Valley, CA 94941,
(415) 381-1750 www.boundarysolutions.com dklein@boundarysolutions.com

Based on 14 years of working on SEAMLESS USA, a 3,140 county open records national parcel layer (NPL), Boundary Solutions, Inc. (BSI) is posting a white paper (link below) to share some thoughts about the NPL's current realities and future prospects. After its first full decade of development, about 85 percent of all U.S. parcels now have a digital parcel map. However, there are emerging problems. Nine hundred Have Map counties are closed records, some charging as much as $1,000 to $300,000 per year for their digital parcel map, restricting use to just a few. With annual report records showing the property tax base of Have Map counties growing far faster than No Map counties, and with nearly every urban county now having a map, increased (accelerated) urban-rural economic disparity is an unwelcomed possibility. Because of the for-profit sponsored digital parcel maps of No Map counties, offered at prices the county and most others cannot afford, the total accumulative benefit of these modern property records falls far short of what is possible when all are using the NPL. All these problems stem from the NPL both not being finished and not being universally open records. Remedy is the following two proposed campaigns: (1) Open Records USA: Earlier annual report record analysis is expanded to 2009 to show that property tax base growth trends observed during the bubble continue (amplified?) during the burst. When shown to GIS managers, they opt to use open records because they are more cost efficient. (2) SEAMLESS USA: Adding a map to all No Map counties is expedited by (1) Recalculating ROI based on increased transaction levels observed in hundreds of comparable counties upon adding an open records map; and (2) Mapping companies recruiting private/NGO interests to fund digitizing of parcel maps they just have to have in exchange for the free use of parcels maps of hundreds of counties they just like to have. The end result is a 3,140 county NPL in place by 2015, not just complete but with open records, to best sustain the NPL's quality, currency, and accessibility needed to maximize its delivery of the greatest good to the greatest many. Follow the link to see the white paper: http://www.npdpdownload.com/SEAMLESSUSANATIONALPACELLAYER

Subject to change as comments are received from other industry leaders involved in this task.

The National Map Partnerships 2, 3:30 p.m. – 5:00 p.m.

U.S. Census Bureau and U.S. Geological Survey Partnership for the Decade
by Andrea Johnson[1] and Dick Vraga[2]

[1]Assistant Division Chief of Geographic Operations, U.S. Census Bureau, andrea.grace.johnson
[2]U.S. Geological Survey, NGP Federal Agency Liaison, 8505 Research Way, MS 511, Middleton, WI 53562, (608) 821-3896 rsvaga@usgs.gov

The U.S. Geological Survey and U.S. Census Bureau have a long history of working together to meet geospatial goals that benefit both agencies and the Nation. With the evolution of *The National Map* and the Census Geographic Support System Initiative, the role and significance of this relationship are growing. This presentation will highlight the goals and structure of the partnership and will explore the benefits to both of the agencies and the geospatial community.

Creating and Utilizing Local-Resolution NHD with LiDAR Data in Florence County, South Carolina
by Dave Arnold,[1] Veronica Moore,[2] and Gary Merrill[3]

[1]U.S. Geological Survey, 1400 Independence Road, MS 645, Rolla, MO 65401 darnold@usgs.gov
[2]GISP, GIS Technician, Florence County, South Carolina 518 South Irby Street Florence, SC 29501, (843) 676-8600 ext. 61431 VMoore@florenceco.org
[3]Geospatial Liaison to South Carolina U.S. Geological Survey - South Carolina Water Science Center, Stephenson Center, Suite 129, 720 Gracern Road, Columbia, SC 29210-7651, (803) 750-6124 glmerrill@usgs.gov

The National Hydrography Dataset (NHD) is a multiple-resolution geodatabase containing hydrography features, including surface-water streams, water bodies, and a number of tables defining relationships and metadata. The hydrographic features are correlated with the certified Watershed Boundary Dataset (WBD). The NHD is maintained and updated via a stewardship process between the U.S. Geological Survey (USGS) and a principal steward located in each State. The USGS coordinates, defines, and distributes the data model. The principal steward is the primary contact for the State and has overall responsibility for what NHD updates are submitted to the USGS. The NHD was originally designed at a scale of 1:100,000 from USGS Digital Line Graphs (DLG) and the U.S. Environmental Protection Agency (USEPA) Reach File Version 3 (RF3). The NHD was later updated to a resolution of 1:24,000; however, in some States the need has arisen to further increase the scale, known as local resolution. The USGS and Florence County, South Carolina, are working together to update the NHD with local-resolution data acquired from Light Detection and Ranging (LiDAR) source data. Placing LiDAR hydrologic breaklines in the NHD provides some specific technical challenges, but once overcome they will allow Florence County to use NHD data at a resolution of 1:4,800. This local-resolution NHD is needed by Florence County to comply with the National Pollutant Discharge Elimination System (NPDES), which is part of the Clean Water Act. The NPDES requires tracking of stream attributes, including flow direction, drainage patterns, perennial and intermittent identifiers, open ditches, and possible outfall locations, all of which will be tracked by using the LiDAR-generated local-resolution NHD.

Assessing Impaired Waters Occurrence Within and Near Federal Lands
by Tatyana DiMascio[1] and Douglas J. Norton[2]

[1]ORISE Fellow, U.S. Environmental Protection Agency, Office of Water, Washington, DC, DiMascio.Tatyana@epa.gov
[2]Senior Environmental Scientist, U.S. Environmental Protection Agency, Office of Water, Washington, DC,
Norton.Douglas@epa.gov

The Federal Government manages public lands consisting of about 30 percent of the 2.27 billion acres of land in the United States. The extent of federal lands and the stewardship responsibilities of federal land management agencies represent a major influence on the condition of the waters of the United States and an opportunity for partnership with the U.S. Environmental Protection Agency (USEPA) in restoring and maintaining U.S. waters. USEPA's Office of Water has partnered with federal land management agencies at the national level to enhance watershed protection, assess restoration needs on federal lands, and mutually aid each agency's statutory programs, strategic plans, and shared mission to protect aquatic resources. The Office of Water carried out joint co-occurrence assessments of the impaired waters occurring within or near public lands and developed interactive tools to help this information become easily used by agencies. Assessments in 2009-2010 include national studies for U.S. Fish and Wildlife lands and U.S. Forest Service lands. Projects for other federal agencies are in planning stages. The U.S. Fish and Wildlife Service assessment is the focus of this presentation. The assessments were carried out primarily through geographic information system (GIS) analysis that utilized the following datasets: (1) The 2002 Impaired Waters Baseline National Geospatial Dataset; (2) The NHD and NHDplus datasets; (3) The Watershed Boundary Dataset (WBD); and (4) property boundaries related to the studied agency. We cross-walked impaired waters data with agency property maps to determine specific properties at risk, the problem waters, impaired causes, and summary statistics, compiled as mapped, tabular, and text products. To promote watershed approaches, the assessment also re-aggregated the project data by 12-digit Hydrologic Unit Code (HUC12) watersheds and reported on the existing total maximum daily loads (TMDLs) for the impaired waters. To enable easy access to the data and results, we developed an internal Web site that contains the final reports and interactive products that allow exploring the results and provide guidance on carrying out a similar study in the future. The Web site elements include briefing materials, GIS datasets, maps of patterns, and spreadsheets of statistics relating impaired waters, federal properties, TMDL reports, and HUC12 watershed boundaries.

The National Map-US Topo, 3:30 p.m. – 5:00 p.m.

The Geospatial PDF: Past, Present, and Future
by Michael Bufkin

Founder and Chief Solutions Architect, TerraGo Technologies, mbufkin@terragotech.com

The Geospatial PDF was first used commercially in 2002 and has become an accepted and popular format for distributing mapping and image data, illustrated by its use as the sole format for the distribution of the U.S. Geological Survey (USGS) quadrangle and US Topo. The format underwent a major bifurcation in 2008, with the introduction of an alternative geo-registration approach by Adobe Systems. While the two methods have been reconciled, there is still some confusion within the public as to their differences and advantages. Now in 2011, TerraGo Technologies has embarked on a series of major enhancements to the methods for creating, interacting with, and viewing of Geospatial PDFs. This presentation will review in detail the history of the invention and development of the format, its current state, and TerraGo's thoughts on future directions for its development, including it use for 3-D display, feature data extensibility, collaborative workflows, and applications delivery. The presentation will include a series of demonstrations of many of these potential capabilities.

Comparison of Topographic Map Designs for Overlay on Orthoimage Backgrounds
by Paulo Raposo[1] and Cynthia A. Brewer[2]

[1]Department of Geography, Penn State University; M.S. candidate, paulo.raposo@psu.edu
[2]Professor of Geography, Department of Geography, Penn State University, cab38@ems.psu.edu

The recently launched 1:24,000 US Topo map series available from the U.S. Geological Survey (USGS) is modernized to include high-resolution orthoimagery. The various vector and image layers in these GeoPDF files may be toggled on and off, enabling a variety of renderings. An initial design for these maps has already been implemented, but the USGS continues to seek effective map designs with improved clarity of content with the orthoimage both present and absent. The research presented here is funded by the USGS to investigate differences in map readability presented by varied ortho-vector map designs. Graphic variables such as color and transparency visually interact with the imagery over which they are drawn. Designs maximizing legibility are most useful to emergency response and natural-resource management personnel who rank among the most common users of topographic maps. The incorporation of imagery into topographic maps complicates the process of choosing effective vector symbols by introducing issues of vertical integration. Orthoimages also offer advantages, including comprehensive update and added detail. Representations of geographic features are different between imagery and vector layers, and an effective map design uses these relationships to actively reinforce and clarify map content. This research experimented with eight different map designs, with visual variables such as color and transparency altered across the orthoimages, vectors, terrain depictions, and labels. Eight different locations were used, from geomorphologically distinct locations around the United States with a mixture of urban and rural settings. For each location, a 5 x 5 inch (12.7 x 12.7 centimeter) extent at 1:24,000 was prepared and printed in each of the eight map designs. Thirty-two participants from the Penn State campus community were solicited to participate in a map reading experiment, divided into eight groups. In a diagram-balanced Latin squares experiment, each group evaluated each of the eight locations once, seeing it drawn in a different design than other groups. Participants answered a set of three scenario-based questions for each map location. After responding to all questions, participants compared the relative ease or difficulty of reading from each map design. The aggregate responses from the eight groups produced a ranking of effectiveness between designs incorporating orthoimages, vectors, terrain, and labels.

Creating the US Topo—A Process Discussion
by Larry Robert Davis

U.S. Geological Survey, 1400 Independence Road, MS 402, Rolla, MO 65401, lrdavis@usgs.gov

In 2009 the U.S. Geological Survey (USGS) began production of the "Digital Map – Beta." The Digital Map – Beta was the first step toward a new generation of digital topographic maps delivered by the USGS. The Digital Map – Beta set the stage for the next step, production of the US Topo. These maps are built from *The National Map* data, which are integrated from local, State, and Federal agencies, along with other sources. Production processes and a variety of software have been implemented to streamline job assignment and tracking, production methods, and product delivery of over 18,000 maps annually. This presentation will discuss the overall US Topo production process and Esri software products, including ArcGIS Server, Job Tracking Extension (JTX), Production Line Tool Set (PLTS), ArcGIS Desktop, and ArcGIS Image Server, which are used throughout the production cycle.

The National Map and LiDAR, 3:30 p.m.– 5:00 p.m.

Accuracy Assessment of a Regional LiDAR DEM
by Roy K. Dokka,[1] Joshua D. Kent,[2] James E. Mitchell,[3] and Kurt L. Johnson[4]

[1]*Director, Center for Geoinformatics, Louisiana State University, rkdokka@c4g.lsu.edu*
[2]*Research Associate, Center for Geoinformatics,, Louisiana State University, jkent4@lsu.edu*
[3]*IT GIS Manager, Louisiana Department of Transportation and Development, jim.mitchell@la.gov*
[4]*Louisiana IT GIS Technical Specialist, Department of Transportation and Development, jim.mitchell@la.gov*

Society has grown dependent on digital elevation models (DEM) to support a wide range of activities, including aspects of planning, engineering design, flood mapping, and so on. Many forget, however, that a DEM is just that, a model, and for any elevation model to be useful, it must be validated with independent observations tied to the National Spatial Reference System (NSRS). The accuracy of almost all digital elevation models is rarely determined. Instead of a proper independent assessment conducted after the creation of the DEM, many accuracy claims are based on the data collection instrument precision estimates, linked to weak or non-existent vertical control. In Louisiana, accuracy assessments have also lacked statistical rigor. To test the validity of the DEM in southeastern Louisiana, including greater New Orleans, an accuracy assessment was conducted using >20,000 measurements from a dual-frequency Global Navigation Satellite System (GNSS) receiver mounted on a vehicle; intrinsic vehicle motions were observed independently. High precision road elevations linked to the NSRS were observed with the Louisiana State University GULFNet real-time positioning system. This system is based on over 50 National Geodetic Survey (NGS) National CORS Network stations. The uncertainty of measurement using the system is +/-5 centimeters (95 percent). Analysis of the DEM shows that the average error (DEM pixel value minus the observed elevation) is near zero, suggesting that the DEM is quite accurate. Although the accuracy assessment shows that much of the DEM is accurate, a spatial analysis shows that the distribution of errors is neither random nor uniformly distributed. Errors are highly correlated with areas where vertical control has been difficult to maintain due to subsidence. The unreliable Light Detection and Ranging (LiDAR)-based DEM has been accepted as best available data and is being used to create new Digital Flood Insurance Rate Maps (DFIRM) and hurricane storm surge models across Louisiana.

North East LiDAR Project: Project Overview
by Michael Shillenn

Vice President Photo Science, 104 S. Church Street, West Chester, PA 19382, (610) 344-0890 mshillenn@photoscience.com

Funded in part by the American Recovery and Reinvestment Act (ARRA) of 2009, as well as direct funding from other Federal, State, and local entities, the North East LiDAR project comprises more than 8,000 square miles of the coastal zone and inland areas spanning six Northeastern States, including Maine, New Hampshire, Massachusetts, Connecticut, Rhode Island, and New York. Photo Science is tasked by the U.S. Geological Survey (USGS) National Geospatial Technical Operations Center (NGTOC) to provide light detection and ranging (LiDAR) data collection and processing, control surveys, product development, and quality control services. This project will not only help stimulate the U.S. economy and provide for more accurate floodplain mapping in the region, but is also a "good government" success story involving multi-state and Federal collaboration that could be used as a model for the national elevation program. The session will include a discussion of the project background and partnerships, supported applications, a technical overview of the project work plan including an innovative approach for developing coastal water elevation models, and a review of project deliverables, schedule, and current status.

Comparison of Airborne LiDAR Elevation Data and USGS National Elevation Dataset Information for Inputs to Regional and Large-Scale Geologic Mapping Applications in Illinois
by Donald E. Luman

Principal Geologist, Office of the Director, Illinois State Geological Survey, 615 E. Peabody Drive, Champaign, IL 61820, dluman@isgs.illinois.edu

As a partner agency in the Illinois Height Modernization Program funded by the National Geodetic Survey, the Illinois State Geological Survey (ISGS) is the primary in-state point of contact for county-based airborne light detection and ranging (LiDAR) elevation data, as well as the responsible agency for providing public access and distribution of all LiDAR data products. The receipt of this high-resolution, improved LiDAR elevation data has afforded ISGS scientists an unusual opportunity to investigate the potential of LiDAR in geographic areas of the State which have been prioritized for geological mapping applications. The airborne LiDAR last return data received by the ISGS averages 1.0-1.2 meter horizontal resolution and 0.3-0.4 meter vertical resolution, and first return data typically exhibit a higher horizontal resolution. This enhanced spatial resolution has resulted in the discovery by ISGS mapping geologists of fine-resolution landscape features within continental glaciated terrain, as well as within areas where bedrock is at or near the surface. The introduction of LiDAR into the ISGS geologic mapping program has created scientific curiosity about the appropriateness and accuracy of the topographic information contained within the U.S. Geological Survey's National Elevation Dataset (NED), most especially the digital elevation models produced from 1 arc-second (30 meter) and 1/3 arc-second (10 meter) source data, both of which are much coarser in horizontal and vertical resolution than LiDAR. Currentness of the NED data is also a concern, particularly in active geomorphic areas of the State, because the source information for the Illinois portion of the NED is several decades old. Using a 10-meter average cell size as the unit of measurement, analysis shows there is a difference of 10 feet between LiDAR and NED elevation values for approximately 44 percent of a typically sized county area; an additional 40 percent of the county area shows little difference. Elevation differences greater than 10 feet are in isolated areas and can be largely attributed to land use change (for example, mining and quarrying), forest cover, and near-surface geomorphic processes. Elevation differences at the 7.5-minute quadrangle scale are more variable and indicate the requirement of LiDAR data for large scale mapping applications.

The National Map Data Themes – Orthoimagery, 3:30 p.m. – 5:00 p.m.

DigitalGlobe's Advanced Ortho Aerial Program
by Michael James Lawless

DigitalGlobe, Inc., Senior Product Manager, mlawless@digitalglobe.com

DigitalGlobe's Advanced Ortho Aerial Program, in partnership with Microsoft, will be the first commercially available wall-to-wall 30-centimeter natural color ortho-imagery program for the Lower 48 States; additionally, this program will create a Western Europe ortho dataset, and also provides full program color infrared (CIR) coverage at 60-centimeter resolution. This efficient and comprehensive program uses a unique camera design, the UltraCam G. Designed for efficient large scale collection, and based on the robust UltraCam technology, the UltraCam G is proving to be an ideal platform for the creation of accurate ortho-imagery at high volume. The flight operation of this program began in mid-2010 and is on track to complete Phase 1 by the end of the second quarter in 2012, at which time the first wall-to-wall set of ortho-imagery products will be complete. Phase 2, a two and a half year follow-on, is expected to refresh the most populous 60 percent of imagery. This program uses 1°x1° blocks as the primary method of organizing collection and production, each of which is approximately 10,000 square kilometers (100 kilometers on a side). Flight programming is based on an efficiency model of collecting these GeoCells as weather and environmental parameters permit, based on north/south zones, keeping the camera fleet as active as possible throughout the year. Our program also provides that each designated metropolitan area is collected leaf off, and processed to even higher specifications within the program. DigitalGlobe believes that this program has the capability to achieve the status of a reference ortho dataset for the Nation, and would appeal to many users of *The National Map*. This presentation will explain the DigitalGlobe Advanced Ortho Aerial Program, including introductions to technology, production, planning, specifications, and availability of this unique and ambitious dataset.

New Jersey US Topo Leaf-off Orthoimage Pilot Project
by Roger Barlow,[1] Dale Benson,[2] and Patrick Meola[3]

[1]*U.S. Geological Survey, 12201 Sunrise Valley Drive, MS 433, Reston,VA 20192, rbarlow@usgs.gov*
[2]*U.S. Geological Survey, National Technical Operations Center, Box 25046, Bldg. 810, MS 510, Denver, CO 80225-0046, dbenson@usgs.gov*
[3]*New Jersey Office of Information Technology, patrick.meola@oit.state.nj.us*

The US Topo7.5-minute mapping series uses an orthoimage base derived from leaf-on National Agriculture Imagery Program (NAIP) orthoimagery as its standard content. Particularly in the eastern area of the country, leaf-on conditions obscure roads, stream, and structures. The U.S. Geological Survey (USGS) and the National Geospatial-Intelligence Agency (NGA) have also invested funds to acquire higher resolution leaf-off orthoimagery that supports State and local missions. The New Jersey

Office of Information Technology working through the USGS Liaison and the National Geospatial Technical Operations Center (NGTOC) offered to undertake a pilot project to mosaic high-resolution orthoimagery as a test case for the New Jersey US Topo image base. The presentation examines issues that occur especially in State bounding quads, and the viability of leaf-off image base for US Topo.

The Wisconsin Regional Orthophotography Consortium—Building a Statewide Partnership
by James P. Lacy[1] and Dick Vraga[2]

[1]*GISP Associate State Cartographer, Wisconsin State Cartographer's Office, 384 Science Hall, 550 N. Park St., Madison, WI 53706-1491, (608) 262-6850 lacy@wisc.edu*
[2]*U.S. Geological Survey, NGP Federal Agency Liaison, 8505 Research Way, MS 511, Middleton, WI 53562, (608) 821-3896 rsvaga@usgs.gov*

The Wisconsin Regional Orthophotography Consortium (WROC) was the focus of significant orthophoto acquisition activity during the spring of 2010. In all, 45 Wisconsin counties and approximately 70 cities, villages, and towns participated in the 2010 WROC project. State and Federal partners contributed over $1.1 million toward WROC projects in 2010. This session will describe the process, benefits, and lessons learned from this multi-year effort.

Listening Session 1, 3:30 p.m. – 5:00 p.m.

Henry Gannett Award Presentation, 5:30 p.m. – 6:00 p.m.

Allison Gannett and Family (Special Guests)

Great-niece of Henry Gannett, Alison Gannett has followed in her Great Uncle's footsteps. She is an adventurous global explorer, World Champion Extreme Skier, Founder of The Save Our Snow Foundation, and an award-winning global cooling consultant. In 2010, Ski Magazine named her "Ski Hero of the Year," while Outside magazine named her "Green All-Star of the Year," next to Leonardo DiCaprio and Arnold Schwarzenegger. An accomplished ski mountaineer with a plethora of worldwide first descents, she also photo-documents climate change as an environmental scientist. Most importantly, she trains individuals, businesses, and governments on her four-step, cost-saving climate change solutions framework, including Al Gore's team, Fortune 500 companies, and businesses around the world, the ski and outdoor industries, school districts, and the U.S. Congress.

Award Reception, 6:00 p.m. – 7:00 p.m.

Conference attendees can meet with Author Frederick Reuss and the other speakers at the conference, and network with friends and colleagues.

Poster Awards, 7:00 p.m. – 8:00 p.m.

Jennifer Sieverling – Enterprise GIS Branch Lead
Barb Ray – USGS Geospatial Liaison for Wyoming

Please join us for the presentation of awards acknowledging work displayed during the joint GIS Workshop and Then National Map Users Conference Poster Session that was held Wednesday evening. Awards will be presented for several categories; some serious and some not-so-serious. Join us in celebrating work well done.

Friday, May 13, 2011, Plenary Sessions, 8:00 a.m. – 9:30 a.m.

Host: Mark DeMulder

Speakers:

Dr. Joel Scheraga

The Importance of Geospatial Information for Effective Adaptation to Climate Change

The climate is changing and will continue to change in the future. The impacts of climate change are already being observed in the United States and its coastal waters, and will continue. Although the scope, severity, and pace of future climate change impacts are difficult to predict, observations and long-term trends indicate that the potential impacts of a changing climate on society and the environment will be significant.

Adaptation is essential. The nation must anticipate and prepare for the risks posed by climate change to human health, the environment, cultural resources, and the economy.

A wide variety of geospatial information will be required to assess site-specific vulnerabilities to climate change and then design and implement effective adaptation strategies (e.g., data and information on climate, water resources, land use and land cover, and topographical information). Using this data, public health officials, resource managers, and other decision makers can project and visualize the possible impacts of climate change, and explore the effectiveness of alternative adaptation strategies. Ultimately, scientists and decision makers must work closely together to ensure the appropriate types of geospatial information are produced (and in a timely fashion) given the particular decisions that are being made.

Larry Sugarbaker

The National Enhanced Elevation Assessment, Preliminary Findings

The USGS is the managing partner for a multi-agency National Digital Elevation Program (NDEP) sponsored assessment of requirements for enhanced elevation data. Information requirements are being identified across Federal, State, Tribal, and other organizations to determine if there is sufficient need to justify a more comprehensive national program. A survey of business needs for LiDAR and IfSAR data, and derived elevation products has identified more than 100 business needs across all levels of government. The requirements are site-specific to national in scope. They include the need to support climate research and monitoring, geologic mapping, urban planning, wind farm siting, precision agriculture, flood hazard mapping, and many other applications. The roles of mapping and geospatial services are expanding to support these increasingly complex activities. The challenge is to build working models that allow us to be responsive to these needs.

Barbara P. Buttenfield

Multiple Representations of Geospatial Data: A Cartographic Search for the Holy Grail?

The demand for a single, unified, detailed spatial database that supports map representations at multiple scales and for multiple purposes continues to challenge many disciplines. This challenge is of particular significance in cartography, where vector features support base mapping for applications in physical and social science across a wide range of scales. The USGS has a mandate to produce geospatial data to standardized specifications, creating databases for different scales and purposes. With emergence of *The National Map,* that mandate is complicated when users expect data layers to be co-registered, fully integrated, and available across a continuous range of mapping scales. For a variety of valid, pragmatic reasons, databases remain isolated, even though it is widely acknowledged that redundant but non-identical versions of geospatial features confound spatial modeling as well as obstruct efficient data management. The presentation will highlight the "multiple representation conundrum," showing selected results from recent Center of Excellence for Geospatial Information Science (CEGIS) work supporting linked multi-scale basemap data management, and identifying challenges to fully linked multiple representations which might be addressed in coming years.

Abstracts

National Hydrography Dataset 1, 10:00 a.m. – 11:30 a.m.

Flow Estimation and the National Hydrography Dataset Plus (NHDPlus) Version 2
by Tommy Dewald,[1] Alan Rea,[2] Kernell Ries,[3] Dave Wolock,[4] Timothy Bondelid,[5] and Cindy McKay[6]

[1]U.S. Environmental Protection Agency, Office of Water, NHD-NHDPlus Program Manager, dewald.tommy@epa.gov
[2]U.S. Geological Survey, Idaho Water Science Center, (208) 387-1323, ahrea@usgs.gov
[3]U.S. Geological Survey, Office of Surface Water, (443) 498-5617, kries@usgs.gov
[4]U.S. Geological Survey, Kansas Water Science Center, (785) 832-3528, dwolock@usgs.gov
[5]Consulting Engineer, Woodville, VA, (703) 987-8592, timothy@trbondelid.com
[6]Horizon Systems Corporation, Herndon, VA, (703) 471-0480, ldm@horizon-systems.com

The NHDPlus is a suite of geospatial products that builds upon and extends the capabilities of the National Hydrography Dataset (NHD) by integrating it with the National Elevation Dataset (NED) and the Watershed Boundary Dataset (WBD). Interest in estimating NHD streamflow volume and velocity to support pollutant fate-and-transport modeling was the driver behind the joint U.S. Environmental Protection Agency (USEPA) and U.S. Geological Survey (USGS) effort to develop NHDPlus, which was first released in late 2006. Widespread positive response to NHDPlus Version 1 prompted the multi-agency NHDPlus team to design an enhanced NHDPlus Version 2 that is currently under production and scheduled for release during mid-2011. NHDPlus Version 2 provides the full characterization of the flow network, identification of unregulated and regulated gages and reaches, and network-based interpolation and adjustment of flows. The Version 2 data model accommodates the ability to specify the percent of water that travels down each path at major divergences as well as water additions, removals, and interbasin transfers. Over 30,000 USGS streamflow gages have been linked to the NHD network and will be used when producing both mean annual and mean monthly streamflow volume and velocity estimates for all networked flowlines in Version 2. These flow estimates will account for the effects of evapotranspiration and are adjusted based upon their network relationships with streamflow gages in the downstream vicinity. These improvements are part of the Enhanced Runoff Method (EROM) used to estimate streamflows with NHDPlus Version 2. Developing this network-based flow interpolation mechanism, and the data to drive it, provides potential for additional improvements to estimate current as well as future streamflows. For example, the framework provided by NHDPlus and EROM could readily be applied to any number of Global Climate Model scenarios to translate those climate-change scenarios into meaningful estimates of local streamflow with an extraordinary level of detail. Such estimates would be very valuable to the water resources community in planning for the future. This conference presentation will provide an overview of NHDPlus Version 2 status and plans for estimating improved streamflows.

Improving the NHD with Diversion Networks
by Kristiana Elite

U.S. Geological Survey, Cartographer, National Geospatial Technical Operations Center, Box 25046, Denver Federal Center, MS 510, Denver, CO 80225-0046, (303) 202-4540 keelite@usgs.gov

The National Hydrography Dataset (NHD) provides a comprehensive representation of the surface water of the United States. These data largely represent the natural flow of water on the landscape using streams, rivers, and lakes. However, in some parts of the country engineered diversion features significantly alter the flow network. Diversion conduits of canals, ditches, pipelines, and tunnels normally present in the NHD have not in the past been properly connected to form the actual flow network. The U.S. Geological Survey is now analyzing and editing the diversion networks to more accurately model flow and make the NHD more suitable for areas where diversions are important. The incorporation of diversions in the NHD will allow for more accurate water modeling, empowering decision-makers and researchers with intelligent information to help address today's water issues and better prepare for the future. This is particularly important in California, where over 2,000 miles of aqueducts, canals, and pipelines significantly affect the flow networks in the State.

Update on Indiana's Local-Resolution NHD Development and the Geo-Synchronization Web-based NHD Maintenance Project
by Phil Worrall[1] and David Nail[2]

[1]*Indiana Geographic Information Council Executive Director, pworrall@igic.org*
[2]*U.S. Geological Survey, 5957 Lakeside Blvd., Indianapolis, IN 46278, (317) 290-3333 ext. 122 dnail@usgs.gov*

In 2008, the State experienced a series of major flooding events that affected 82 of 92 Indiana counties. Three different Federal disaster declarations resulted in estimated total damages over $1 billion. As a result of the 2008 flooding, the State of Indiana has been allocated over $371 million for emergency disaster assistance to fund several programs beginning in 2010-2011. Of this amount, $13.4 million will be used to acquire various GIS-based elevation/hydrologic/hydrographic data and derivative products in support of disaster-mitigation planning, including a continuation of Indiana's Geographic Information Council's Local-Resolution National Hydrography Dataset (NHD) project. In addition, the Indiana Geographic Information Council, in partnership with the U.S. Geological Survey, has had an active Waters Workgroup committee that has been planning the creation, development, preparation and stewardship for a local-resolution NHD layer. This presentation will provide an update on Indiana's local-resolution NHD development and the Geo-Synchronization Web-based NHD maintenance project.

Data Integration, 10:00 a.m. – 11:30 a.m.

Challenges in Integrating *The National Map* Themes of Hydrography and Elevation
by Samantha T. Arundel-Murin

U.S. Geological Survey, 1400 Independence Road, MS 644, Rolla, MO 65401, sarundel-murin@usgs.gov

One of the first large-scale integration projects of two *The National Map* themes has occurred during the digital contour production process for the US Topo. To produce contours for this product, hypsography in the form of the 1/3 arc-second National Elevation Dataset (NED), which is mostly based on 10-meter resolution digital elevation models (DEMs), is forced to conform to specific hydrographic features in, typically, the 1:24,000-scale NHD dataset. Specifically, streams, double-line streams, and waterbodies each require a different treatment. Challenges in integrating the NED and NHD datasets are fairly straightforward, even though the datasets have been developed on different time scales. They include contour stream re-entrant realignments, water body avoidance, and interpretive double-line stream crossings. Re-entrants change due to stream channel meandering, divergence, redirection, or obstruction. Water bodies swell, recede, evaporate, and emerge (typically manmade). Double-line stream channels vary greatly over time, but their general characteristics exist already in the 10-meter NED. Recently both the NED and NHD programs have undertaken practices to increase the accuracy of their datasets, resulting essentially in a higher resolution dataset for each theme (for example, Light Detection and Ranging (LiDAR)-based NED and 1:4,800-scale NHD). Hence, while hydrographic/hypsographic integration involves only two datasets, these datasets currently have at least four different integration scenarios: low-resolution NED/low-resolution NHD, low resolution-NED/high-resolution NHD, high-resolution NED/low-resolution NHD, high-resolution NED/high-resolution NHD. The complexity of the integration process varies with each scenario, with additional obstacles existing in each case. Some of these challenges will be presented here.

Issues in the Concurrent Integration of NHD and NED Updates
by Ricardo Lopez-Torrijos,[1] Erika Boghici,[2] Lucinda McKay,[3] and Cheryl Rose[4]

[1]*Institute for the Application of Geospatial Technology, Senior GIS Analyst, Ricardo.Lopez.inAlbany@gmail.com*
[2]*TerraCarto LLC., erika.boghici@terracarto.com*
[3]*Horizon Systems Inc., ldm@horizon-systems.com*
[4]*crose002@gmail.com*

A project covering the New York City (NYC) water supply watershed aims to provide its water quality and watershed protection programs concurrent updates of National Elevation Dataset (NED) and National Hydrography Dataset/ Watershed Boundary Dataset (NHD/WBD), at resolutions meaningful for activities from best-practice implementation to modeling and adaptive management, all integrated into the NHDPlus architecture. We present some of the issues encountered in the process. Compilation of these Light Detection and Ranging (LiDAR) and aerial photography updates requires the adoption of data capture and processing standards supporting the regulatory, management, and modeling needs of NYC, mindful of the characteristics of

LiDAR collections, and providing for correct vertical NHD and NED alignment. This last requirement leads to the adoption of the terrain data as the primary reference. Compilation protocols need to address the many situations due to ground cover, LiDAR point cloud variation and water connectivity situations, elevation conflation methods, as well as compliance with the NHD data model. Compiled NHD linework is used to clean the LiDAR point cloud and improve its classification. Conversely, geomorphic derivatives of the LiDAR data are developed at several points of the process to inform the linework compilation, provide a linkage to reservoir catchment hydrologic characteristics, and ensure vertical alignment of the NHD and NED products. The ability to develop a three-dimensional hydrographic network raises new requirements for NHD quality control and database loads. The final output in the NHDPlus schema will allow integration of user business data with the water cartographic representation, use of COTS tools in modeling applications, and will extend these tools at lower costs by leveraging community modeling efforts.

Canadian-U.S. Hydrographic Data Harmonization and Integration
by Michael T. Laitta

International Joint Commission of Canada and the U.S., Physical Science Advisor, Physical Scientist, GIS Coordinator, 2410 Pennsylvania Ave. NW, Washington, DC 20037, (202) 736-9022 (202)341-1487 laittam@washington.ijc.org

The International Joint Commission (IJC), in coordination with Environment Canada, Natural Resources Canada, U.S. Geological Survey, Agriculture and Agri-Foods Canada, has made substantial progress with the harmonization of the shared fundamental hydrographic datasets along the Canadian-U.S. interface. Phases I and II of this effort, the alignment and editing of sub-drainage areas within the major Transboundary Basins and the first pass connection of the fundamental hydrographic layers (the U.S. National Hydrography Dataset [NHD] and the Canadian National Hydro Dataset [NHN]), are close to completion. In 2011, this effort will focus on the delineation and refinement of smaller drainage units within these now-harmonized sub-drainage areas—the extension of the U.S. Watershed Boundary Dataset (WBD) into Canadian territory. Coincidental to this next step, the IJC is encouraging the development/expansion of binational water quality and quantity applications such as StreamStats and the U.S. SPARROW model. This presentation will touch upon the basic technical methods employed to facilitate the negotiation of binational delineations, impacts to the Federal stewarding agencies, and potential opportunities for sustainable, regionally based yet binational hydrologic applications for the shared U.S.- Canadian geographic interface.

The National Map and Other Data 1, 10:00 a.m. – 11:30 a.m.

Coordination of Activities of the NHD, NHDPlus, WBD, and StreamStats Programs
by Kernell Ries,[1] Tommy Dewald,[2] Karen Hanson,[3] and Jeff Simley[4]

[1]*U.S. Geological Survey, Office of Surface Water, 5522 Research Park Drive, Baltimore, MD 21228, kries@usgs.gov*
[2]*U.S. Environmental Protection Agency, Office of Water, 202-566-1178, Dewald.Tommy@epamail.epa.gov*
[3]*U.S. Geological Survey, Utah Water Science Center, 2329 W. Orton Circle, West Valley City, UT 84119, (801) 908-5038, khanson@usgs.gov*
[4]*U.S. Geological Survey, Box 25046, Denver Federal Center, MS 510, Denver, CO 80225, (303) 202-4131 jdsimley@usgs.gov*

Three themes of *The National Map* (TNM), the National Hydrography Dataset (NHD), the Watershed Boundary Dataset (WBD), and the National Elevation Dataset (NED), are used in the development of the NHDPlus dataset and the StreamStats Web application. The NHD provides a digital representation of surface water for the Nation at scales of 1:100,000 and 1:24,000, with data at larger scales available in some areas. Network-navigation functionality built into the NHD allows analysis of movement of water, sediment, and chemical constituents along the stream network, and upstream and downstream relations among activities along streams. The WBD is a representation of drainage boundaries for the Nation at six nested levels. The WBD boundaries were determined for specific points along the stream network from delineations based on hydrography and topography at a scale of 1:24,000 in the conterminous United States. The NED is a seamless raster of elevation points along the land surface, primarily derived from U.S. Geological Survey (USGS) 10- and 30-meter Digital Elevation Models (DEMs) but increasingly derived from higher resolution sources, such as Light Detection and Ranging (LiDAR). The NHDPlus dataset was developed for use in hydrologic and water-quality applications. It includes nine layers that were derived from the 1:100,000-scale NHD, the WBD, the 30-meter NED, and the USGS National Land-Cover Dataset (NLCD). NHDPlus includes delineations of the land area (catchment) that contributes streamflow to each NHD stream reach, and associated attributes such as stream slope and catchment area are available to characterize each stream reach and catchment. StreamStats is a GIS-based Web application that is used primary to delineate drainage boundaries for user-selected sites and to obtain estimates of streamflow statistics, such as the mean flow and the 100-year instantaneous peak flow, for gaged and ungaged sites on streams. In addition, StreamStats includes numerous tools that rely on stream-network navigation. Drainage-area delineations in StreamStats rely on datasets prepared individually for each State from the NHD, NED, WBD, and NHDPlus. The NHD, WBD, NHDPlus, and StreamStats development teams are working closely together to maximize efficiency in the development, utility, and quality of data. This presentation will include brief descriptions of each dataset and application, along with discussions of ongoing and planned coordination activities. A panel discussion following the presentations will allow the audience to ask questions of the presenters.

Integrating Kentucky Karst Data into the National Hydrography Dataset (NHD)
by James Seay,[1] Robert Blair, P.G.,[2] Deven Carigan,[3] Phil O'Dell, P.G.,[4] Joseph Ray, P.G.,[5] and James C. Currens, P.G.[6]

[1]Kentucky Division of Water, Geoprocessing Specialist II, james.seay@ky.gov
[2]Geologist, Registered, Kentucky Division of Water, Robert.Blair@ky.gov
[3]Geologist II, Kentucky Division of Water, Deven.Carigan@ky.gov
[4]Groundwater Hydrologist II, Kentucky Division of Water, Phillip.O'Dell@ky.gov
[5]Geologist Registered (retired), Kentucky Division of Water, sat.jar@att.net
[6]Hydrogeologist, Kentucky Geological Survey, currens@uky.edu

Karst groundwater drainage is prevalent throughout the Commonwealth of Kentucky. Approximately half of Kentucky is underlain by soluble rocks with karst development potential. Nearly 25 percent of the State has significant karst groundwater drainage. Surface water and groundwater are conjunctive systems, and this connection is very direct in karst areas. Groundwater quantity and quality have a strong influence on surface drainage in karst regions. The Kentucky Geological Survey (KGS) and the Kentucky Division of Water (KDOW) have compiled and digitized karst flow data for much of Kentucky and published these data as the Kentucky Karst Atlas map series. The Karst Atlas was developed to spatially document karst data obtained by numerous authors. The U.S. Geological Survey (USGS) and KDOW funded a pilot study in 2010 to determine the feasibility of integrating karst data with the National Hydrography Dataset (NHD). The West Fork Red River watershed (0513020606) was selected as the study area, due to the presence of a wide variety of karst features and the quality of the available digitized data. Data from the Karst Atlas were entered into the NHD, using the NHD Geo Edit toolset. Groundwater flowlines were categorized with the Underground Conduit FType; whereas, the site locations were categorized as either SinkRise™ or SpringSeep™. The presentation will cover various issues that were encountered while updating the NHD, ranging from bugs in the NHD Geo Edit tool developed by the USGS (resolved in the current version) to features missing from the NHD (submitted to the USGS using their new NHD Issues reporting system).

The National Map Applications 1 and 2, 10:00 a.m. – 11:30 a.m.

Vegetation Characterization Data made available as WMS Utilizing *The National Map*
by M.P. Mulligan and Tim Mancuso

U.S. Geological Survey, Box 25046, Denver Federal Center, MS 302, Denver, CO 80225
mpmull@usgs.gov, tmancuso@usgs.gov

Biological Informatics Program (BIP), Vegetation Characterization Program (VCP), web serves highly detailed large scale (1:5,000 - 1:25,000) vegetation geospatial information datasets of U.S. National Park units. Eventually there will be about 280 National Park units represented. Working with *The National Map* (TNM) development team BIP will be deploying a Web Map Service (WMS) for these datasets through the newly developed TNM application (target 1st quarter 2011). Initially about 50 park units will be represented. The work will implement within the BIP's National Biological Information Infrastructure (NBII) portal one of TNM development's targeted user stories. The scenario involves a user in the NBII portal at the VCP online data location wishing to inspect a view of the data before downloading it. Utilizing TNM Viewer button, now added to the page, the user will have access to a WMS version of that dataset for assessment as to its extent and applicability to their needs, perhaps in context of other additional germane TNM data layers. For instance, if the user is in mrdata (http://mrdata.usgs.gov), and the user wants to show a mrdata WMS in TNM Viewer, the user clicks on a new "open in TNM Viewer" link on mrdata site (which was auto-generated by mrdata site-generation capability), which tells the URI call to TNM what the WMS is, and auto-opens the TNM viewer, zoomed to scale/spot with mrdata WMS turned on.

The National Map Viewer Base Map and Services
by Calvin Meyer
U.S. Geological Survey, 1400 Independence Road, MS 506, Rolla, MO 65401, clmeyer@usgs.gov

The U.S. Geological Survey (USGS) has modernized visualization and download capabilities to improve user experience of *The National Map* (TNM). Managed by the USGS National Geospatial Program, TNM has transitioned its assets and viewer application to a new visualization and delivery environment that includes interoperable base maps, integrated download services, and an improved viewer platform.

The National Atlas of the United States 1:1,000,000-Scale Hydrography Dataset: An Overview of the Dataset, the Production Process, and New Mapping Products
by Florence E. Thompson

U.S. Geological Survey, Geographer, fethomps@usgs.gov

The National Atlas of the United States is producing a 1:1,000,000-scale hydrography dataset that consists of streams, streamgages, water bodies, and coastlines of the United States, Puerto Rico, and the U.S. Virgin Islands. Hydrographic features were selected, generalized, and refined from the 1:100,000-scale National Hydrography Dataset to create regional networked geodatabases and nationwide shapefiles that serve multiple cartographic and scientific purposes. The hydrographic features are vertically integrated with other 1:1,000,000-scale National Atlas cartographic frameworks such as transportation features and national boundaries. The harmonization of U.S. hydrographic data with 1:1,000,000-scale streams, water bodies, and coastlines from Canada and Mexico supports mapping and data analysis at continental and global scales through the North American Atlas and the Global Map. This presentation will provide an overview of the contents of the dataset, a description of the production process, the applications of the dataset, and the incorporation of the dataset into National Atlas mapping products.

Management and Population of Sites: A National Water Information System Application for Automatically Populating Sitefile Information
by Steven K. Predmore[1] and Scott B. Whitaker[2]

[1]*U.S. Geological Survey, 4165 Spruance Road, Suite 200, San Diego, CA 92101, (619) 225-6153 spredmore@usgs.gov*
[2]*U.S. Geological Survey, 1400 Independence Road, MS 602, Rolla, MO 65401, (573) 308-3516 swhit@usgs.gov*

The Management and Population of Sites (MAPS) application is a collaborative effort between the National Water Information System (NWIS) and the National Geospatial Technical Operations Center (NGTOC). The application provides two major functions: (1) a map-based graphical user interface (GUI) for entering and editing sitefile information into NWIS and (2) a Web service, GeoPoint Query Service (GPQS), to query geospatial data from *The National Map* and other sites for automatically populating available geospatial site data. The map-based GUI consists of a map form, which is a static display for site locations, as well as data entry and edit forms for manual entry of data. The base map is the standard U.S. topographic map, but any layers a user has may be added. The GPQS is designed to take latitude, longitude, and horizontal datum entered manually or from a click on the map form and return geospatial data from *The National Map*, the National Atlas of the United States, and GeoCommunicator, which is from the U.S. Bureau of Land Management and U.S. Forest Service.

An Interactive, GIS-based Application to Estimate Continuous, Unimpacted Daily Streamflow at Ungaged Locations in the Connecticut River Basin
by Peter A. Steeves and Stacey A. Archfield

U.S. Geological Survey, 10 Bearfoot Road, Northborough, MA 01532, psteeves@usgs.gov, sarch@usgs.gov

Daily time series of streamflow are critical for solving any number of hydrologic problems. Because most stream reaches are ungaged, these data are commonly needed for rivers that have no readily available measurements of streamflow. In the Connecticut River Basin, dam operations and their effects on the aquatic habitat are of particular interest. Here, daily time series of streamflow are needed as input to reservoir simulation and optimization models of the Connecticut River Basin as well as for developing ecological-flow prescriptions for rivers and streams in the Connecticut River Basin. For this reason, the U.S. Geological Survey (USGS), in cooperation with the Northeast Association of Fish and Wildlife Agencies and The Nature Conservancy, has developed a map-based point-and-click tool, termed the Connecticut River Basin Sustainable Yield Estimator (SYE) to estimate a daily streamflow time series at ungaged sites in the basin. Daily streamflow at the ungaged location is estimated by solving a series of regression equations that relate streamflow magnitudes to measurable basin attributes. An index streamgage is used to assemble the regression-estimated streamflows into a time series. The Connecticut River SYE tool is built on the USGS StreamStats platform and spans the entire basin, including the States of Connecticut, Massachusetts, New Hampshire, and Vermont. Users are able to point and click on a stream location of interest in the Connecticut River Basin and obtain a delineated watershed, basin attributes, and a daily time series of streamflow.

Use of *The National Map* at FEMA
by Doug Bausch[1] and Sara Brush[2]

[1]*FEMA, Senior Physical Scientist, Douglas.Bausch@dhs.gov*
[2]*FEMA Region VIII, Mitigation Denver Federal Center, Bldg 710a Denver, CO 80225, (303) 235-4871 Sara.brush@dhs.gov*

The Federal Emergency Management Agency (FEMA) is a major user of *The National Map* products for our Risk MAP (Mapping, Assessment, and Planning) program that produces flood hazard maps for the Nation and works with communities to apply these data to make communities safer. To support this mission and enhance product quality, Risk MAP is making a major investment in elevation data. These data will make a major contribution to the elevation component of *The National Map*. FEMA also uses *The National Map* and U.S. Geological Survey (USGS) products to support applications for disaster response and recovery. In addition, USGS products are used to drive our preparedness exercises, including the May 16th, 2011, New Madrid Earthquake National Level Exercise (NLE). The presentation will provide an overview of FEMA, the Risk MAP program, and a number of applications of *The National Map* to support FEMA's mission.

Why *The National Map*? 10:00 a.m. – 11:30 a.m.

Making the Transition from Paper to Digital Maps
by James E. Mitchell[1] and Sean Deinert[2]

[1]*IT GIS Manager, Louisiana Department of Transportation and Development, jim.mitchell@la.gov*
[2]*GDM International Services, sdeinert@gdmis.com*

Over the past 40 years, geographic information systems (GIS) technology has caused a great paradigm shift in the collection, management, analysis, and production of map data. As the geospatial revolution began, maps were the primary source of geospatial data. It was common to digitize data from paper or mylar maps and create GIS data from them. As advances in technology made computers more efficient, faster, and able to store more data, the model shifted from "the maps make the data" to "the data make the maps." Regardless of its format (digital or paper), a map is a model. That is, it is an abstraction of reality designed to represent the geophysical characteristics of a locale. This fact remains constant, despite the substantial differences between digital geospatial data and static map presentations. Both forms of geospatial data provide advantages and disadvantages to the user. This presentation outlines the similarities and dissimilarities between digital and paper maps. These differences provide challenges to users, as well as to those who produce, manage, and distribute geospatial data. Developing an appreciation for the distinctions between digital, geospatial data and static maps will provide a better understanding of how both play an important role in *The National Map*.

The National Map: Why Bother?
by Jay B. Parrish

Professor of Practice, Dutton Institute, Pennsylvania State University, jbp3@psu.edu

The National Map (TNM), viewed as a set of ready-made products and services for regular consumption by end users, is underutilized after almost a decade of construction. It is less well known in the user community than Google Earth or Bing Maps and likely lacks the user base of government alternatives such as the National Aeronautics and Space Administration's (NASA) World Wind or the U.S. Geological Survey's (USGS) own Earth Explorer. This paper, however, is an argument in favor of TNM. To favor its continuation and development, one has to look at TNM as something other than the way it is defined above. The USGS has the same job in geography that it has performed for more than a century: to provide base-mapping data for the country. In the 21st century, though, the means of delivering the data have changed radically. It used to be that the USGS produced series of graphic maps, authoritative maps with sound, reasonably current data embedded in them. Much of what the USGS used to do in terms of making maps to deliver data is now being done in the private sector. The need for authoritative data, though, persists. Gathering and maintaining data from which maps can be made is an unglamorous chore, with no end and no reward, is hard to explain to customers, and harder to explain to those who might support the task with funding. The USGS needs to continue to create and (or) aggregate data in some way and to publish it to the user community. The original topographic map series was a base product which was used by a multitude of entrepreneurs, sometimes even just reselling the product with a mark-up, yet the net result was the economic development of a Nation for a century. Would yesterday's topographic program have passed the ROI analyses that exist today? TNM is in the same position of power to transform.

Standards and Specifications for *The National Map*
by Kristin A. Fishburn

U.S. Geological Survey, National Geospatial Technical Operations Center, Box 25046, MS 510, Denver Federal Center, Denver, CO 80225, (303) 202-4405 kafishburn@usgs.gov

Standards and specifications are essential to facilitate the development and sharing of geospatial data and products. The U.S. Geological Survey (USGS) National Geospatial Technical Operations Center (NGTOC) has rejuvenated its internal standards activities during the past year. Projects for FY 2011 include the development of standards and specifications for the US Topo, *The National Map* Viewer, Scanned Historical Maps, National Boundaries Data, Orthoimagery, and Elevation Data. NGTOC

Standards Specialists will address standards and specifications for all thematic data included in the *The National Map* over the longer term. NGTOC is pursuing formal publication for new and revised documents and is coordinating its work with external standards organizations via the USGS external standards liaison. Standards Specialists are also developing review and approval processes and cycles, updated document templates, and a modernized USGS standards Web site.

The National Map Data Themes – Names, 10:00 a.m. – 11:30 a.m.

Using the Geographic Names Information System for Interagency Consistency: How the U.S. Census Bureau Has Integrated the USGS Federal Identification Codes into Its Database
by Michael R. Fournier

U.S. Census Bureau, Geographer, michael.r.fournier@census.gov

Beginning with the migration of place codes from the Federal Information Processing Standards (FIPS-55) to the American National Standards Institute (ANSI), the U.S. Census Bureau has worked with the Board on Geographic Names (BGN) to incorporate the Geographic Names Information System (GNIS) names and codes into the Bureau's Master Address File/Topologically Integrated Geographic Encoding and Referencing (MAF/TIGER) geographic database. This presentation will focus on the process undertaken to coordinate and implement a process by which the U.S. Geological Survey (USGS) and the Census Bureau have worked together to establish a names and codes system which can be used by all government agencies and maintain the consistency needed to minimize cost and maximize utility. It will describe how, in 2007/2008, the USGS provided dBASE GNIS files to the Census Bureau and how the Census Bureau matched the GNIS files against the geographic entities common to both files to transfer GNISIDs to Census Bureau records and achieve consistency with common data attributes. Since then there has been continuous updating and periodic additional matching to maintain consistency. It will also review a program in which the Census Bureau utilized point landmarks from the GNIS to assist Census enumerators in wayfinding and how, as a result, the Census Bureau is also assessing the use of the GNIS as a potential source for improving the coverage of landmark features it uses on its map products on an ongoing basis.

The Changing Role of the Geographic Names Information System: Feature Locations
by Corey Plank

U.S. Bureau of Land Management, Lead Cartographer, corey_plank@or.blm.gov

From topographic quadrangles to the Geographic Names Information System (GNIS) to *The National Map*, geographic names traveled from paper maps to tables to electronic database and back to electronic and paper maps. Along the way, some of the names have changed. More importantly, locational precision often dropped, as coordinates were measured for inclusion in the GNIS. GNIS is the storehouse of official names of geographic features. In its electronic format, it became a source for locational analysis and selection. Numerous Web sites utilize the positions in GNIS to point users to nearby attractions. Errors exist in GNIS. Coordinates measured from maps of various scales, typos, and blunders in capturing features created many of the misplaced features currently held in GNIS. A unique partnership in the State of Oregon identifies and submits for correction GNIS points with bad locations or spellings, duplicated information, and missing data. This presentation covers the causes of errors in GNIS, reasons that they should be corrected, and the process used to identify those errors. We will discuss the evolution of GNIS from a list of official names to a vital source of information for dozens of Web sites and mapping applications.

Collection of Critical Structures and Facilities Data for *The National Map*
by Ellen Currier, Maggie Drews, and Chrissy Fellows

GIS Analysts, Lane Council of Governments, ecurrier@lcog.org, mdrews@lcog.org, cfellows@lcog.org

The U.S. Geological Survey (USGS) entered into a multi-year partnership with the Geospatial Enterprise Office of the Oregon Department of Administrative Services (DAS-GEO) and the Lane Council of Governments (LCOG), to promote and coordinate the development of landmark and structure data within the Oregon Spatial Data Framework, and to leverage State efforts in order to improve related Federal datasets, specifically the Geographic Names Information System (GNIS) and the National

Structures Dataset (NSD). Project goals included identification and documentation of available data sources for 33 high priority structures types for all 36 counties in Oregon. Beginning with the 13 coastal counties in Oregon and later extending to the remaining counties in the State, LCOG staff is currently compiling detailed information on 33 different types of structures. Some of the structure types collected include: Hydroelectric Facilities, Airport Terminals, Oil and Gas Facilities, Road and Railroad Tunnels, Colleges and Universities, and Wastewater Treatment Plant. A parallel task to develop a Stewardship Plan and identify potential stewards for landmark and structure data was completed in 2010. The two tasks helped inform and improve each aspect of the project. Throughout the 3-year span of this project, the project teams shared the experience gained through data collection, outreach, and networking resources. Upon completion of the project, the USGS and State of Oregon will receive updated, consistent, and high priority structures data for all 36 Oregon counties. The data will be ready for inclusion in the National Structures Database and *The National Map*. The data will comply with the NSD standards and be ready for inclusion in appropriate Oregon Spatial Data Framework themes. An immediate benefit of this project is a more complete and up-to-date representation of critical structures and facilities in State, Federal and local databases. A longer term benefit comes from LCOG's statewide data search and inventory of structures data sources. The inventory will provide useful input, especially the development of specific maintenance procedures. The inventory will also help focus Federal-State partnership efforts around important data sensitivity and maintenance issues associated with sharing and stewardship of these critical datasets.

Mashathon, 12:45 p.m. – 2:15 p.m.

Flood Inundation Mapping Layers and *The National Map*
by David Nail

U.S. Geological Survey, 5957 Lakeside Blvd., Indianapolis, IN 46278, dnail@usgs.gov

The U.S. Geological Survey (USGS) is developing a Flood Inundation Mapping Initiative (FIMI) to meet USGS science strategy goals for the National Hazards, Risk, and Resilience Assessment Program. The goal of this initiative is to develop a tool for flood response and mitigation using digital geospatial flood-inundation maps that show flood inundation and flood-water depths on the land surface. These mapping data, produced as vector shapefiles and as raster depth grids, will be made available to the National Weather Service, as input to FEMA's HAZUS software, to emergency managers, and to the public as a Web-mapping service and offered as downloadable files. This presentation demonstrates the use of *The National Map* Viewer as a means to display and to provide access to the download service for the vector shapefiles and raster depth grids. In addition, this presentation proposes that these data become components of standardized data themes of *The National Map*.

The National Map Partnerships 3, 12:45 p.m. – 2:15 p.m.

From the Ground to the Cloud, from Maps to Web Services: The U.S. Forest Service Experience
by Susan J. DeLost,[1] Betsy Kanalley,[2] Barry Napier,[3] and Aaron Stanford[4]

[1]*U.S. Forest Service, Geospatial Management Office, National Geospatial Services Program Manager,*
(703) 605-4578 sdelost@fs.fed.us
[2]*U.S. Forest Service, Geospatial Management Office, National Geospatial Services Assistant Program Manager,*
(703) 605-4575 bkanalley@fs.fed.us
[3]*Geospatial Management Office - Geospatial Service and Technology Center (GSTC) Director, Geospatial Service*
and Technology Center, (801) 975-3498 bnapier@fs.fed.us
[4]*Geospatial Management Office - Geospatial Service and Technology Center GIS Analyst, (801) 975-3806 astanford@fs.fed.us*

The U.S. Forest Service has been collecting data and producing maps since its inception in 1905. Much of that work has been accomplished in collaboration with the U.S. Geological Survey and other partners. We will briefly chart the agency's legacy of mapping, the present state of mapping, and migration to an enterprise data center environment, which provides data and maps as Web services. We will look into the future as we move toward the cloud computing environment, all as we carry out our mission of "Caring for the land and serving people" during the International Year of Forests.

Updating Names in the NHD in Oregon: Putting Stewardship to the Test
by Robert Harmon[1] and Meredith Carine[2]

[1]Oregon Water Resources Department, GIS Coordinator, robert.c.harmon@wrd.state.or.us
[2]Oregon Water Resources Department, meredith.j.carine@wrd.state.or.us

In 2010 the Oregon Water Resources Department (OWRD) received funding through a U.S. Geological Survey (USGS) Partnership Cooperative Agreement to update National Hydrography Dataset (NHD) features with missing names from the Geographic Names Information System (GNIS). More than 6,000 GNIS NHD features do not have an official name, for a variety of reasons. These features are distributed widely throughout many governmental jurisdictions and comprise an array of types, error conditions, and solutions which will be illustrated during the presentation. There were two major drivers for this effort. The first is that this project will integrate NHD and the GNIS datasets, which will provide direct benefit to *The National Map* and the US Topo. The second is that the Pacific Northwest Hydrography (PNWH) NHD stewardship process will be fully implemented and refined as discrepancies between the two datasets are resolved. This project serves as a pilot for other States, and the presentation will cover the issues, processes, and methodologies developed during the project.

Geologic Community of Use (CoU)
by Michael Cooley,[1] Greg Allord,[2] Dave Greenlee,[3] and Kent Brown[4]

[1]U.S. Geological Survey, 12201 Sunrise Valley Drive, MS 511, Reston, VA 20192, mcooley@usgs.gov
[2]U.S. Geological Survey, 465 Science Drive, Suite A, Madison, WI 53711, gjallord@usgs.gov
[3]U.S. Geological Survey, 47914 252nd Street, MS SAB, Sioux Falls, SD 57198, greenlee@usgs.gov
[4]Utah State Geological Survey, kentbrown@utah.gov

The U.S. Geological Survey's (USGS) National Cooperative Geologic Mapping Program and its State Geological Survey partners have a need for 1:24,000-scale topographic base-map data, cartographically symbolized, delivered with positional accuracy that supports the compilation and publication of geologic maps for the Nation. In 2010, the National Geospatial Program's (NGP) *The National Map* (TNM) along with members of the National Cooperative Geologic Mapping Program formalized a pilot effort to identify and document the requirements as well as determine what NGP can provide that meets these requirements. This requirements document was used by NGP to pilot several prototype maps that demonstrate what can be accomplished today along with an informal report that explains how the maps were made. A subsequent phase will test the procedures by having the geologic mapping partners make a production map using the NGP data. A final phase for this year's activity is to document the next steps to be worked on by NGP so that base data can be more efficiently used by the National Cooperative Geologic Mapping Program. The Seminar will include (1) the goals of the Geologic Community of Use (CoU); (2) an overview of the geologic requirements needed for the base map; (3) an example of a base map generated from TNM data as defined by the geologic requirements document; (4) a report that outlines the process on how to initiate a geologic base map with TNM data; and (5) a brief synopsis on what works, what does not, and what needs development. A follow-on demonstration will be conducted to show both the pilot maps made by NGP based on current capabilities and the final maps by the geologic mapping community. This session will demonstrate how the data are used in a geologic map production process, along with a candid assessment of what works, as well as the challenges that remain.

Team: Randy Orndorff and Ernest Allen Crider, USGS National Cooperative Geologic Mapping Program; Kent Brown, Utah State Geological Survey; Greg Allord, Stafford Binder (retired), Michael Cooley, Rob Dollison, David Greenlee, and Jim Barrett (USGS contractor), USGS National Geospatial Program

The National Map and Other Data 2 and 3, 12:45 p.m. – 2:15 p.m.

Integrating Regional Biodiversity Occurrences with *The National Map* Topographic Services—A Test Case for Mobilizing and Visualizing U.S. Geological Survey Science Data Using Distributed Services
by Derek Masaki,[1] Mark Wimer,[2] Rob Dollison,[3] and Joe Miller[4]

[1]USGS NBII Pacific Basin Information Node Technical Coordinator, dmasaki@usgs.gov
[2]USGS Patuxent Wildlife Research Center, Laurel, MD, mwimer@usgs.gov
[3]USGS National Geospatial Program, Reston, VA, rdollison@usgs.gov
[4]USGS National Biological Information Infrastructure, Program Office, Reston, VA

Species occurrence data are essential for taxonomic and biodiversity research and are invaluable for mapping historic and predicted distributions of species and as an educational tool. Without a spatial context and a means to relate occurrence data visually within a broader geographic dimension, additional analysis and interpretation are problematic. This demonstration provides an example of how multiple data services from U.S. Geological Survey (USGS) units can be merged to provide an enhanced information resource. This collaborative effort between the USGS National Biological Information Infrastructure (NBII), Patuxent Wildlife Research Center (PWRC) and *The National Map* is working to demonstrate data integration methods that merge and overlay several spatial layers from distinct sources. The effort leverages the extensive observation collections and taxonomic expertise of PWRC along with the geospatial technical capacity of *The National Map* and the NBII. The resulting map service brings together species occurrence elements, high-resolution aerial imagery, and ecological attributes (climate, soil types) to provide users with a visual data interface for query and navigation. In addition to the spatial location, attribute data cross-link to species taxonomy and natural history through Web references.

Stream View Concept
by Alex K. (Sandy) Williamson

USGS (retired) and University of Washington, Geology Instructor, P.O. Box 1271, Spanaway, WA 98387,
cell: (253) 376-8273 H: (253) 531-1481 Fax: (815) 346-3397 sandeb2@gmail.com

This idea is for a Web site and database of photos, still, 3-D, and video, oriented entirely to streams, like Google's Street View, but for streams. Each entry would be stream indexed to the National Hydrography Dataset (NHD) and geolocated in space and time. There would be a date selection slider to view a date range of photos on the mapped area of interest. Uses for these photos, mosaics, 3-D views, and videos along streams are envisioned to be:

1. Public visualization of streams everywhere as easily as they can visualize streets now—more attention leads to more and better conservation.

2. Virtual travel for those working with streams to the stream reach[es] in question without leaving their desks to investigate anomalies, do reconnaissance, and so on. This would have a wide audience among fishermen, kayakers, and rafters.

3. Documentation of riparian habitat conditions and changes from now or in the past to the future and visualization of extent of stream changes projected from the future back to present or some other time.

4. Easy visualization of stream channel and riparian area conditions and effects.

5. Public access to historic U.S. Geological Survey (USGS) stream photos (taken over last 140 years) in a systematic way.

6. Vacation planning with more connection to rivers—maybe partnering with Park Service.

7. Documentation of erosion and sedimentation of banks and bars along streams.

8. Documentation of land use changes near streams.

9. Inventorying of geomorphic features in stream channels.

10. Provision of a frame of reference about flooding issues and what a flood depth means visually.

11. Provision of mash-ups with other applications that are geographically tied to streams.

There would be a unified map-driven user interface to (1) sets of single stills and maybe videos and (2) 3-D views of some stream reaches. Group 1 photos would be shown as clickable points on a map and shown as thumbnails to the side, and Group 2 would be shown as colored highlighting along stream reach[es] showing availability of 3-D views. Sources of photos:

Existing sources, like Google Picasa and Yahoo Flickr using an application programming interface to subset (and NHD stream index) pictures that are currently geotagged to be near streams (after seeking help and (or) permission). Maybe in the future there will be place recognition software for water features and (or) landform features, like there is face recognition software now.

Partners like Flickr or Picasa/Panoramio to host a repository of user-contributed stream photos, from stream-watch groups and government and others: USGS photos – historic photos at 20,000 stream gages over the last 130 years, volunteer-taken photos, repository of metadata and links to outsourced photos , and 3-D views of streams taken by 3-D camera setups along a stream reach. The USGS and others could help facilitate projects of photo documentation using inexpensive ($1,500) 360-degree camera setups.

Anticipating Plausible Environmental and Related Health Concerns Associated with Future Disasters
by Geoff Plumlee,[1] Suzette Morman,[2] Charlie Alpers,[3] Todd Hoefen,[4] and Greg Meeker[5]

[1]*U.S. Geological Survey, Box 25046, Denver Federal Center, MS 973, Denver, CO 80225, gplumlee@usgs.gov*

[2]*U.S. Geological Survey, Box 25046, Denver Federal Center, MS 964, Denver, CO 80225, smorman@usgs.gov*

[3]*U.S. Geological Survey, Placer Hall, 6000 J Street, Sacramento, CA 95819, cnalpers@usgs,gov*

[4]*U.S. Geological Survey, Box 25046, Denver Federal Center, MS 964, Denver, CO 80225, thoefen@usgs.gov*

[5]*U.S. Geological Survey, Box 25046, Denver Federal Center, MS 973, Denver, CO 80225, gmeeker@usgs.gov*

Disasters commonly pose immediate threats to human safety but can also produce hazardous materials (HM) that pose short- and long-term environmental-health threats. The U.S. Geological Survey (USGS) has helped assess potential environmental health characteristics of HM produced by various natural and anthropogenic disasters, such as the 2001 World Trade Center collapse, 2005 Hurricanes Katrina and Rita, 2007-2009 southern California wildfires, various volcanic eruptions, and others. Building upon experience gained from these responses, we are now developing methods to anticipate plausible environmental and health implications of the 2008 Great Southern California ShakeOut scenario (which modeled the impacts of a 7.8 magnitude earthquake on the southern San Andreas fault, http://urbanearth.gps.caltech.edu/scenario08/), and the recent ARkStorm scenario (Atmospheric River 1,000, modeling the impacts of a major, weeks-long winter storm hitting nearly all of California, http://urbanearth.gps.caltech.edu/winter-storm/). Environmental-health impacts of various past earthquakes and extreme storms are first used to identify plausible impacts that could be associated with the disaster scenarios. Substantial insights can then be gleaned using a Geographic Information Systems (GIS) approach to link ShakeOut and ARkStorm effects maps with data extracted from diverse national- and State-scale database sources containing geologic, hazards, and environmental information. This type of analysis helps constrain where potential geogenic (natural) and anthropogenic sources of HM (and their likely types of contaminants or pathogens) fall within areas of predicted ShakeOut-related shaking, firestorms, and landslides, and predicted ARkStorm-related precipitation, flooding, and winds. Because of uncertainties in the event models and many uncertainties in the databases used (for example, incorrect location information or lack of detailed information on specific facilities) this approach should only be considered as the first of multiple steps toward a more quantitative, predictive approach to understanding the potential sources, types, environmental behavior, and health implications of HM predicted to result from these disaster scenarios. Although only a first step, this qualitative approach will help enhance planning for, mitigation of, and resilience to environmental-health consequences of future disasters. This qualitative approach also requires careful communication to stakeholders that does not sensationalize or overstate potential problems but rather conveys plausible impacts and next steps to improve understanding of potential risks and their mitigation.

The National Map Science Support Application
by Lance Clampitt

U.S. Geological Survey, Rocky Mountain Area, Regional Science Office, 2327 University Way, Suite 2, Bozeman, MT 59715, lsclampitt@usgs.gov

The National Map serves as a foundation for integrating, sharing, and using spatial data consistently while retaining and improving other valued characteristics, such as positional accuracy and national consistency. *The National Map* Viewer provides a gateway to robust *National Map* datasets and serves a multitude of users across the Nation. Currently *The National Map* Viewer may not be the first choice or best solution for strategic geospatial science support within the U.S. Geological Survey (USGS). The proposed *National Map* Science Support Application builds on the tools found in *The National Map* Viewer and utilizes *The National Map* data as the backbone for a customized utility built to specifically support USGS science needs. The goal of *The National Map* Science Support Application is to package a geospatial application that easily interacts with and supports USGS science. This application customizes and simplifies the geospatial tools required by a majority of USGS scientists. The application focuses on the integration and visualization of project-specific geospatial data with *The National Map* data as well as partner data sources to provide the best geospatial support for USGS science collaboration.

Mapping Coastal Wetlands in the National Hydrography Dataset—The Louisiana Experience
by Sean Deinert,[1] James E. Mitchell,[2] and Kurt L. Johnson[3]

[1]LADOTD/GDM International Services, sdeinert@gdmis.com
[2]IT GIS Manager, Louisiana Department of Transportation and Development, jim.mitchell@la.gov
[3]IT GIS Technical Specialist, Louisiana Department of Transportation and Development, kurt.johnson@la.gov

Coastal wetlands are among the most dynamic landscapes on Earth. In Louisiana, this is further complicated by the presence of the Mississippi River Delta, which has been characterized as one of the most active deltaic regions of the world. These facts combine to create a great challenge in geospatial data development and mapping along the Louisiana coast. U.S. Geological Survey (USGS) 1:24,000-scale topographic maps in this region average two decades since their last revision. Some of these maps were last produced a century ago. At the same time, the National Oceanic and Atmospheric Administration's (NOAA) nautical and coastline charts represent data collected before 1940 along the northern Gulf coast. Bathymetry along the Louisiana coast is likewise outdated, and much of it dated back into the 1800s. As a whole, existing geospatial data are not sufficiently up to date to use as maps or geographic information system (GIS) data for analysis. The National Hydrography Dataset (NHD) represents the water features from USGS maps and, as digital data, also includes a flow model to route water through the hydrologic systems of the United States. However, the flow model assumes a classical downhill hydrologic regime that follows the terrain and elevation changes. Furthermore, coastal basins are often distributary networks. The NHD flow network is modeled after a typical hierarchy of basins increasing in stream order. Coastal areas violate these assumptions by experiencing regular tidal effects, as well as occasional storm events that cause flow to reverse and travel upstream against the elevation gradient. This presentation will examine the challenges of photorevising coastal watersheds in the NHD. The unique distinctions between coastal and classical watersheds will be reviewed and how these distinctions led to the development of the procedure for coastal watershed revision developed by the Louisiana Department of Transportation and Development, for the USGS.

National Fish Habitat Action Plan Data Delivery
by Andrea C. Ostroff

U.S. Geological Survey, Fisheries and Aquatic Resources Biological Informatics, National Center, MS 302,
12201 Sunrise Valley Drive; Reston, VA 20192, (703) 648-4070 aostroff@usgs.gov

The National Fish Habitat Action Plan (NFHAP) is aligning a great number of stakeholders including Federal, State, non-governmental, and private partners to address the need for broad-scale conservation action focusing on issues of declining fish populations, health, and habitat. The backbone of this effort is a science-based assessment analyzing large amounts of existing and new information to facilitate the implementation of the NFHAP. Because the NFHAP is science-based, there is a great need to integrate data from many sources to assist meeting the plan's goals. The NFHAP Science and Data Team is developing a data system that will facilitate the data management and data exchange that supports the implementation of the NFHAP. What is currently available is a map-based data viewer to broadly disseminate the national inland and coastal fish habitat assessment results to partners and stakeholders. The assessment results report the relative risk of fish habitat degradation based upon available national datasets that were analyzed in a comparable fashion across the United States. The hierarchical spatial framework based upon the National Hydrography Dataset Plus supports data summary, analysis, and visualization at multiple spatial scales with the goal of providing tools to decision-makers at multiple policy levels. It is anticipated that through the regional partnerships new data will be collected and integrated into future analyses. These data in addition to the national assessment data can be made available through *The National Map*, through which the potential for new opportunities and innovative approaches can be identified and realized.

Listening Session 2, 12:45 p.m. – 2:15 p.m.

Closing Session – What You Said: Shaping the Direction of *The National Map*, 2:15 p.m. – 3:00 p.m.

Please join us in reflecting on what the community said about *The National Map* during the conference, and how the National Geospatial Program will use this information.